D1218595

Site Selection: New Advancements in Methods and Technology

"Thompson Associates has evaluated every site we have considered since our inception and has helped us in our ongoing expansion efforts."

Jack A. Smith
President and CEO
The Sports Authority

"The reputation of Thompson Associates in the site location research and sales forecasting industry gave us confidence to expand the Chick–fil–A freestanding concept. Thompson Associates' performance over the last three years has only enhanced our respect for their expertise. We can ill afford to make site decisions without their input."

Bureon Ledbetter
Vice President Real Estate and General Counsel
Chick–fil–A

Site Selection: New Advancements in Methods and Technology

Second Edition

Robert W. Buckner

Lebhar-Friedman Books
Chain Store Publishing Corp.
A Subsidiary of Lebhar-Friedman, Inc.
New York

Printed in the United States of America.

ISBN 0-86730-717-X

Contents

Page Figures, Maps and Tables

Credits

Contributing Authors	Peter C. Davis David L. Huntoon Robert F. Kennedy Dr. James D. Root James A. Swanson Mark E. Zygmontowicz
Editor	Marcia Mogelonsky
Project Director	Sue Moskowitz
Cover Design	Mary Jo Scibetta
Proof Reader	Laura Glenn

Acknowledgements

The management of Thompson Associates would like to extend its gratitude to a number of people that have been involved in the completion of this book. First, we would like to acknowledge the efforts of Dr. Robert Swartz, Professor of Geography, as a contributing editor in our first draft of the book. Bob's knowledge of the industry and his academic perspective provided guidance in making Site Selection: New Advancements in Methods and Technology, 2nd Edition, an instructional book for the retail and educational communities.

We would also like to extend our thanks to Mary Zahner and Naydeen Stamats at Thompson Associates, for their patience and persistence in helping us complete several transcripts as well as the final copy. Finally, many thanks to the Thompson Associates Staff for their assistance in helping us create a book that does justice to John Thompson's original text on this subject.

Preface

Retailers make few decisions that are as permanent (and potentially unforgiving) as selecting store locations. In their book Site Selection: New Advancements in Methods and Technology, the owners of Thompson Associates provide the reader with an honest, pragmatic assessment of both the merits and the shortcomings associated with the science and art of store location research. The techniques and methodologies addressed in this book have real-world applicability. As such, it can be used as a "how to" manual by individuals who are becoming acquainted with the store location research process, as well as seasoned veterans.

Site selection and location research have, historically, been synonymous with estimating sales for new retail stores; and the value of accurately assessing these sales is obvious. However, the field of store location research has evolved dramatically over the past several years to the point that the information derived during the development of forecasting methodologies is terribly under-utilized if it is used only to forecast sales. At Home Depot, we use this information: 1) to screen markets so that we can identify those with the greatest potential; 2) to strategize markets in order to ensure optimal store deployment; 3) to review our store network in existing markets to identify opportunities for additional stores; and 4) to look into the future to assess implications of competitive and demographic changes within markets. These recent innovations, as well as the more traditional sales forecasting applications, have been addressed throughout this book.

The perspectives presented in this book have been based on practical, real-world experiences. Undoubtedly, many of the methodologies and techniques disscussed herein reflect the experiences derived from the longstanding relationship between our two firms. John Thompson and his colleagues started working with The Home Depot in 1979 when we were in the concept-stage of development. Since then, we have been both the beneficiaries of, and the real-word laboratory for, their location research efforts. However, Home Depot is but one of the over 1,000 clients that Thompson Associates have served over the past four decades. The reader is fortunate to be able to share these real-world experiences in a book of this nature.

Arthur M. Blank
President/CEO
The Home Depot

Bryan J. Fields
Senior VP–Real Estate
The Home Depot

1 Store Location Research: Background and Scope

IN TODAY'S COMPETITIVE RETAIL ENVIRONMENT, SALES and market share preservation and growth are critical to survival. In this context, it is not surprising that most successful retailers are constantly engaged in self-evaluation: How can we serve our customer better? How can we more effectively manage the business? and so on. An important component of this self-evaluation process is an ongoing assessment of a retailer's store-deployment plans (Where should the next store be located? Which new markets provide the greatest sales opportunities? Does the deployment of existing stores adequately serve our consumers?). More often than not, complacent retailers who do not routinely evaluate their current locations and assess future deployment options, find themselves overcome by rapidly expanding, aggressive competitors.

A location-related error is permanent. While merchandising-, pricing-, marketing-, and management-related problems can be addressed in a relatively short time frame, once a site is selected and a store is built, there is little that the retailer can do. The options include continued operation of an unprofitable store (a less tenable option when faced with numerous locational problems), closing the store to minimize losses, or the relocation of the store. Because the costs associated with poor location decisions are severe and unforgiving, most successful retailers thoroughly research the viability of the sites they consider before making a long-term commitment.

Location Research Defined

Historically, location research focused primarily on assessing the viability of a specific site but the process has expanded well beyond its historical application. "Location research" is currently used to identify and quantify factors that most significantly affect store sales performance, as well as to develop and apply tools which aid in the identification of potential expansion opportunities, and current store deployments. Ultimately, the role of location research is to facilitate the formulation of a logical real estate strategy that addresses a retailer's unique location requirements while maximizing its sales potential and market share.

The location research process is used by a retailer to evaluate issues which are complementary and consistent with the goals of the corporate real estate strategy. For example, an aggressive retail chain that intends to expand into new markets in the future must first determine which markets offer the greatest potential. Typically, this is accomplished by conducting market screenings, during which the viability of individual markets (at a regional or national level) is assessed. After a retailer has identified the markets that would appear to provide the best expansion opportunities, a detailed store-deployment strategy is completed for each market to determine the actual number of new stores that can be supported, the optimal store deployment, and an estimate of the sales potential for the new market as a whole. If the results of the strategy indicate that the market offers a favorable opportunity for new store development, the process of evaluating the sales potential for individual sites that comprise the deployment strategy is conducted.

Location research is also used by retailers to assess their current store positioning within existing markets. Such an evaluation provides the basis for identifying potential opportunities for in-fill stores, as well as store relocation and remodel/expansion candidates within each existing market. Further, consumer research is often conducted in existing markets to identify opportunities for improvements in store operations, merchandising, and marketing. The goals of assessing existing markets in terms of both store deployment and consumer research are better positioned stores, more responsive operations, and better-directed marketing, all of which ultimately enhance sales performance and chain market share.

History of Location Research

The challenge associated with selecting the optimal location(s) in a market is not new, although it is now considerably more complicated because of the increasingly dynamic retail environment. Originally, site selection consisted of the retailer visiting a potential site ("kicking the dirt"), driving the area, and reaching a "gut feel" decision based on his experience and knowledge of a particular market. Often the "gut feel" approach proved adequate in helping the retailer find viable locations. The reliability of this approach diminished, however, as competition increased and the dynamics of the marketplace changed. Retailers soon came to the conclusion that the site selection process required a more systematic and sophisticated methodology.

Quantitative site selection techniques were pioneered in 1932 by William Applebaum, who is widely recognized as the "father" of modern store location research.[1] Applebaum endeavored to address systematically the pragmatic challenge of estimating the sales potential of proposed sites for new stores. His approach focused on the study of existing retail stores in which customers were interviewed at the store to determine where they lived. Using customer residences, Applebaum was able to define the primary trade areas for a sample of stores. This information, in conjunction with store sales levels, provided a basis for evaluating the sales potential of future locations. Applebaum reasoned that if the estimated trade area of a potential location had an analogous number of people and competitors, then it was likely that comparable sales levels could be achieved at the proposed store.

This technique of using analogies and inferences derived from studying the performance of existing stores became known as the analog method of sales forecasting. It is a powerful technique which remains in use today, although with far more sophistication in its application of demographic and lifestyle characteristics, competition, and related market and location dynamics in the forecasting process.

Applebaum and others attempted to modify the basic analog methodology by adding variables and detail to subsequent analyses. More than 30 years ago, Economic Geography, devoted a special issue to the topic of store location research. Several of the articles in this issue attempted to identify the basic ingredients associated with site selection analyses and successful retail locations. The factors identified included population, income levels, competition, access, economic considerations, and trade areas. Intuitively, these factors would appear to have some bearing on the sales potential associated with a proposed retail site. We have subsequently learned, however, that a simple set of standard variables, while appropriate for some retailer concepts, may be inadequate for others. Often, the very assumptions which were incorporated in the analog process to enhance its accuracy, resulted in misleading conclusions simply because the variables assumed to be related to sales potential, did not really affect a store's sales performance level.

Since Applebaum, many researchers have worked to develop more advanced site evaluation systems. These developments include gravity models, regression models, normal curves, and spatial allocation and interaction models, as well as advancements in the development and application of the analog system.

Methodological Innovations

The analog method has evolved into a widely adaptable forecasting tool as a result of the integration of statistical techniques into its development process. More specifically, analog development today uses regression and correlation analysis to quantify the core customer profile of a retailer more effectively, and to accurately measure the impact of competition and other market factors affecting sales performance. The knowledge gained from this "enhanced" approach to analog development has resulted in not only greater forecast accuracy, but has also yielded information that benefits the merchandising and marketing programs of a retail chain. Enhanced analogs are a quantifiable procedure that relies on a systematic application; they are flexible enough to allow subjective input from a trained analyst. But, the use of enhanced analogs requires an investment in time and money on the part of the retailer.

An alternative to the analog method is the gravity model. Based on Reilly's law of retail gravitation[2] (i.e., the probability of a consumer choosing a specific store is a function of store size and distance to the store), the gravity model is widely used today in the supermarket and drugstore industry. The development of the gravity model as a forecasting tool is largely credited to Dr. David Huff. In the late 1960s, Dr. Huff's model was modified and applied by SuperValu (a wholesale and retail grocer located in Minneapolis, Minnesota) to estimate the sales potential of proposed company-owned and independent supermarkets. Since the development of the SuperValu gravity model, several additional gravity simulation models have been developed and are widely used throughout the supermarket industry.

While the development of gravity models was based on an analysis of a relatively large sample of existing supermarkets, gravity models are generic "off the shelf" site evaluation tools which can be used to produce sales forecasts for most supermarkets, and some other types of convenience-oriented retailers such as drugstores. Unlike analog forecasting systems, which require a significant up-front expenditure of effort and money to develop, gravity models can be applied immediately by a trained research analyst. To apply gravity models, the analyst must be able to estimate accurately the sales volume of competitive stores, and trade area expenditure potential (a measure of the total dollars spent within a trade area on relevant merchandise categories). Such information is usually only easily obtainable for a few retail segments such as supermarkets and drugstores. Further, because gravity model sales projections are heavily influenced by store size and distance traveled, but largely ignore the demographic/psychographic makeup of trade areas, the use of gravity models is practical only for convenience-and commodity-oriented retailers.

The 1960s also witnessed the introduction of regression-based sales forecasting models. The John S. Thompson Company was one of the first to develop and apply regression models as a tool in the retail site evaluation process on behalf of its retail clients. Today, Thompson Associates continues to use regression and other forecasting approaches (among which include enhanced analogs and gravity); the specific approach used is dictated by the needs and requirements unique to each individual retailer. Regression models were also developed and used internally by many retail chains to evaluate the sales potential of new sites. These regression models were based on multiple regression analysis, a statistical technique that seeks to select the combination of variables that most accurately predicts potential sales. Like the development of analogs, regression models are based on an analysis of the performance of existing stores. When completed, the regression model is represented by a mathematical equation (or equations) requiring the incorporation of the values of the predictive variables.

An advantage of a regression model is that it does not necessarily require an experienced analyst and it is relatively easy to apply. The primary disadvantage is that a regression model in its purest form is essentially a "black box" approach to forecasting—that is, it does not have the flexibility to incorporate the qualitative or subjective analysis that is often necessary when considering the unique characteristics of a particular location or market. Further, it is limited by the variables in the regression equation, which, in unique or extreme circumstances may result in an inaccurate forecast.

There are several other less commonly used approaches to the site evaluation and selection process. A system often referred to as normal curves was developed at Safeway supermarkets in the early 1950s. Normal curves quantify the relationship that exists between a store or chain's sales penetration levels and the distance a customer travels to shop a store. Basically, normal curves graphically depict the rate that a retailer's average per-capita sales decline as distance from a store location increases. While normal curves are relatively simple, they have proven to be effective when used as a forecast verification procedure in concert with another site evaluation system.

A more recent methodological development is the use of spatial allocation models. Spatial allocation models have seen limited use as a forecasting methodology in the past because of the intense data requirements associated with this approach. As a result of efficiencies gained from increasingly more powerful computers and machine readable databases, however, the use of spatial allocation models to forecast retail sales may become more practical. A spatial allocation model systematically evaluates a number of potential locations in a specific market to identify those sites that maximize potential sales by simultaneously examining supply and demand relationship. The technique allocates the total potential sales for a chain within a market to each of the individual store sites thereby providing sales estimates for each location. While spatial allocation models are in theory an appropriate approach for evaluating store networks, they have not been practically applied in the retail industry. This is primarily because spatial allocation models are costly to develop and their application requires the modeling of an entire market (as opposed to the portions of a market that are of particular interest to a retailer), which greatly expands the time and expense necessary to perform the analysis.

Ultimately, a retailer who is interested in approaching the real estate decision-making process from a more quantitative perspective must decide which of the various techniques or methodologies is most appropriate for them. In part, the answer to this question depends on whether the retailer simply requires a means of improving the "odds" that the sites ultimately developed will be successful, or whether he requires a system that is capable of generating a sales forecast upon which a pro forma analysis for the site can be based. The answer to this question is a function of the nature of the retailer (convenience or destination-oriented, for example), as well as the number of stores that could be included within a database. As noted previously, the methodology that is effective in predicting potential sales for a supermarket may not be appropriate when evaluating the potential for a discount department store, a home improvement store, a mall-based retailer, or a restaurant. Faced with a variety of techniques and methodologies, the researcher must decide which approach is most applicable given his objectives and the limitations or constraints implicit in the database.

The use of several site evaluation methods can minimize the weaknesses inherent in any one approach, thereby resulting in a more accurate estimate of a site's sales potential. For example, the rigidity inherent in regression-based models can be mitigated by a complementary forecast verification procedure using an analog system which incorporates the subjective input of an analyst. These independent evaluations enhance the reliability of the findings.

Even the most well-conceived site evaluation system is of little value if the retail concept "misses the mark" with respect to consumer needs and expectations. In this context, it is not surprising that consumer research is routinely used to assess perceptions regarding a retailer's pricing, selection, service, quality of merchandise, shopping environment, advertising, and related strategic issues, all of which significantly effect a retailer's performance. Consumer research projects can be conducted at a variety of levels for an equally varied number of reasons. For example, consumer research conducted within the trade areas of individual stores can help determine why a store is under performing. On a market or regional basis consumer research may be used to assess current market conditions. A review of the retail industry's most successful operators reveals a common bond; these retailers consistently monitor consumers to take advantage of changes occurring in the marketplace. By periodically surveying the consumer, the retailer is able to develop merchandising, operational, and marketing programs that specifically address the attitudes, needs, and opinions of the consumers being served by their stores.

---

1. William Applebaum (ed.), Store Location and Development Studies (Worcester, MA: Clark University, 1962); ibid., A Guide to Store Location Research (Reading, MA: Addison Wesley, 1968).
2. William J. Reilly, 1929, as reported in Richard L. Nelson, The Selection of Retail Locations (F.W. Dodge Corp., 1958), pp. 148–149.

2 Data Commonly Used in Market Research — Types and Sources

IN DETERMINING WHERE TO OPEN NEW STORES, IT IS important to recognize the difference between the terms location and site. "Location" is a marketing term, normally used in referencing the characteristics (e.g., market potential, population density, demographic characteristics, competitive environment, trade area accessibility, retail synergy, and so on) of a general area, which may (or may not) represent a viable site for a retailer. "Site," on the other hand, is a real estate term referring to a physical parcel of land; the characteristics of a site include, among other factors, size and shape of the parcel, its visibility, ingress/egress, parking (capacity and configuration) and layout. Hence, it is entirely possible that a good site might be a terrible location, or an excellent location might be a mediocre site. For example, there are an almost unlimited number of good sites in the middle of Kansas—numerous sites that are large, relatively flat and square, with two major frontage roads (each with multiple ingress/egress points), all having excellent visibility, and plenty of parking. However, these sites represent terrible locations for most retailers, since they are in areas of very little population density, and thus no appreciate market potential.

Whether evaluating sites or assessing retail locations, all market-research activities involve the gathering, analysis, and interpretation of information. Such information may be very general (economic characteristics), or extremely specific (the number of prospective customers who reside in a particular section of a trade area). Information to be collected and analyzed may involve specific measurements of density (the number of people in a trade area), percentages or ratios (the estimated market share of a competitor), characteristics of a trade area (consumer income levels), or more qualitative observations (the ease of entering or leaving a shopping center parking lot, or the traffic flow and congestion near a site). It may consist of primary data (information that is gathered first hand by the research analyst), or it may be secondary data (data that has already been gathered by another source and then re-used by the research analyst). Regardless of the type, knowing what information to gather, where to find it, and how to evaluate its accuracy are integral parts of the research process.

The accuracy of conclusions derived from any research investigation is directly dependent upon the accuracy of the underlying data. The often-used computer phrase "garbage in, garbage out" is particularly relevant to a discussion of data; it is not very practical to expect good results from an analysis utilizing bad or questionable data. The likelihood that inferences and conclusions derived from the analysis are accurate, is significantly enhanced if the data adequately represent the situation at hand. But if the accuracy of underlying data is suspect, if the data are not representative, if the data are not gathered in an unbiased manner, or if the data are inconsistent or inconclusive, the research may produce results that do not answer the original query in a valid way.

In discussing data and their usefulness, statisticians typically refer to two types of characteristics: the statistical validity of the data, and statistical reliability of the results derived from analysis of the data. Validity is defined as a situation in which data are "well-grounded or justifiable, having a conclusion that is directly drawn from premises or inferences." In this sense, the validity of data refers to how accurate or representative it is. Reliability, on the other hand, is defined as "the extent to which an experiment, test, or measuring procedure yields the same results on repeated trials." Broadly interpreted, statistical reliability means the extent to which the same research methodology, using the same data, produces the same results repeatedly. Therefore, it is the researcher's task to evaluate all data used in a research investigation, as to its validity (accuracy) and reliability (consistency).

There are several types of data of vital concern to the location-research analyst. These include measures of density (population, households, or businesses), demographic characteristics (the demographic composition of the density components being analyzed), competitive information, site characteristics, character of the market, and consumer expenditure potential.

Population Measurements and Demographic Characteristics. One of the most important pieces of information used in retail location research involves measurements of the potential number of customers. Generally, density and demographic information is collected for the area within which a store's influence is felt, and from which it obtains the majority of its customers and sales. It is in this area (referred to as the store's trade area) that the study of population measurements and demographic characteristics is most important. The specific measurement to be evaluated differs by retail type and the nature of the retail product sold. For example, an appliance store would likely be more concerned with the number of households (rather than persons) in an area; most households have only one washing machine regardless of the number of persons residing within the household. The more relevant measure for a supermarket would be the population in the area; after all, everyone has to eat.

While measures of the number of potential customers in the absolute sense is important, the nature of the customers is often as significant, especially to a store whose image transcends just locational convenience. Certain demographic characteristics (such as income, education, occupation type, age, household size, or type of domicile) often have a significant impact on the probability of success for a retail store—especially one with a specific appeal to a demographically defined segment of the population. Just as a home center may be concerned with the number of owner-occupied houses in its trade areas (since renters typically invest far less in home maintenance and repairs), an upscale women's apparel store may be concerned with the presence of high income levels in its trade areas (since lower income consumers typically would not shop at such a store); just as a toy store may be concerned with the number of households with children under age 12 in its trade areas (since its merchandise is oriented toward young children), a computer superstore is more likely concerned with consumers having higher incomes and educational levels (since less well-educated persons are not as likely to purchase computers and related merchandise).

The Bureau of the Census (a division of the U.S. Department of Commerce) is the primary source of population and demographic data in the United States. Comparable governmental agencies exist in almost every county—for example, Statistics Canada (Statscan) is responsible for collecting demographic information in that country. The U.S. Census of Population and Housing, conducted every ten years, produces the most accurate and complete demographic information available. While some of this information is based on tabulations of questionnaires that have been administered to virtually every household in the country, there is additional data that results from more extensive questionnaires that were administered to a smaller sample of households. The resulting data are then presented for the country as a whole, individual states, counties, metropolitan areas, cities, towns, and all the way down to the census tract, block group or individual block levels.

At each level of geography, different types of demographic information are available. Generally speaking, the larger the geographic area for which the data are published, the more data are available; less data are available for smaller areas, in order to preserve the confidentiality of people residing in such smaller areas, and depending upon the method of sampling. In other words, the Census Bureau is most sensitive to revealing data from which individual information can be obtained; where the number of individuals or households in a geographic area is so small as to enable inference of individual household characteristics, the data are typically not reported. Where such suppression occurs, it obviously impacts the statistical validity and reliability of the resulting data.

The most frequently used types of data available from the U.S. Census of Population and Housing include the following:

- Population
- Number of Households
- Household Size
- Age Groups
- Income Levels
- Employment Types
- Educational Attainment
- Type of Residence—single-family versus multi family, owner-occupied versus renter

While the U.S. Bureau of the Census generates the most comprehensive data on a wide range of geographic levels, a significant weakness lies in the fact that this information is collected, tabulated, analyzed and published only once every ten years. For many parts of the United States (i.e., where growth or decline is slow, consistent, and predictable), this does not pose much of a problem to the location research analyst. But for other areas (where growth may be dramatic and sporadic), the availability of demographic data only once a decade can result in an analyst drawing inaccurate inferences, missing trends or incorrectly calculating market potential.

To address this problem, a number of demographic research firms have come into being during the 1980s and 1990s. These firms typically use as their base data the most recent decennial census. Then, using available local, regional and national data, they construct computer models to generate estimates of current population and demographic characteristics for intercensal years, as well as projections of future population and demography. Such data are usually made available to market researchers in one of two ways: on an annual licensing basis, in which the licensee is equipped with annually updated computer disks or CD-ROMS, from which all demographic data for all geographies are available; and on a site-by-site basis, based on requests of data users.

Much of this data tends to be relatively consistent from vendor to vendor, especially for areas that are not explosive in their growth characteristics. This is due to the fact that, not only do all vendors use as their base the decennial census data, but they also tend to use many of the same local, regional and national sources for their update models. There are, however, instances in which this consistency is lacking. For example, one data vendor may do a more thorough job of researching local sources of income data, thus resulting in better income estimates; another vendor may do a superior job of estimating median age levels. While the research analyst could evaluate each data vendor accordingly, in order to select the set of data that are most reliable for the purposes at hand, it is probably more important to use one consistent approach and to gain significant insight into the strengths and weaknesses of only one national dataset.

In high-growth areas, the accuracy and reliability of data obtained from vendors is generally questionable. This is due primarily to the inability of computer models to portray accurately what is actually happening on a local level in such markets. Here, the analyst would do well to gather and analyze whatever local density and demographic information he can obtain, in order to minimize errors in estimating market potential resulting from incorrect data. Local sources for such up-to-date information include public utilities (such as telephone, electric, gas and water companies), municipal departments (such as local planning/zoning departments, building inspectors, or department of streets and sewers), and local agencies (such as industrial development departments, or Chambers of Commerce).

Competitive Data. Another data component important to a location research investigation concerns the competitive environment within which a prospective retail store will operate. There is little doubt that the presence or absence of relevant competition plays an important role in assessing the sales potential for a new store. Further, a market area that appears understored at the present time may not be so in the future; it is unlikely that a site being considered by the location analyst is the only new store planned for the trade area.

There are several sources for gathering competitive information, the most accurate of which is firsthand data collected through field work on the part of the analyst. Field work enables not only the identification of competitive locations, but also the gathering of competitive characteristics as well. Quantitative and qualitative assessments such as competitive store sizes, location types, shopping center types and cotenants, merchandise lines and emphasis, operational characteristics, or store conditions can be observed and measured by the analyst during an on-site visit. Many firms involved with location research and site analysis develop and utilize standardized reporting forms for gathering competitive data, in order to ensure that the same type of information is gathered in a consistent fashion from market to market, study to study, and time period to time period, resulting in data reliability and consistency (refer to Figure 2-1 for an example of a competitor information card).

FIGURE 2-1
COMPETITOR INFORMATION CARD

Store Name ______________________________

Street ______________ **Cross Street** ______________

City & State ______________ **Quad:** NE SE SW NW

Store Type: Conventional ______________ Superstore ______________

Department ______________________________

Gross Area (sq. ft.) __________ **Sales Area** (sq. ft.) __________

Hours: Monday–Friday __________ Saturday __________ Sunday __________

Number of Checkouts: Regular ______________ Express ______________

Location Type:

Freestanding _____ Neighborhood _____ Community _____ Regional _____

Signage: Pylon _______ Monument _______ Store Front _______

Parking Spaces ______________ **Ingress/Egress Points** ______________

Shopping Center Name ______________________________

Cotenants ______________________________

Competitive information can also be obtained from numerous secondary sources. One of the leading sources for competitive data is the telephone book, particularly the business listings by category (this information is now available on CD-ROM). In using such a source, it is important to review not only the major category of competitive retailers, but also related competitive categories. For example, when looking for all relevant competition for natural food stores, the research analysis would want to access not only the Grocery Store category, but also the categories for Health Food Stores and Vitamin Stores. Similarly, when inventorying competitors for a sporting goods store, the researcher would want to look not only at the Sporting Goods Retailers category, but also at the categories for Athletic Footwear, Team Sports, Department Stores and Discount Department Stores. We caution, however, that because the retail environment is ever changing, competitive information obtained from the telephone books can be quite dated.

Other sources of competitive information are the directories published by trade associations, Chambers of Commerce, and other types of directory publishers. One such source of retail information is Chain Store Guide (published by CSG Information Services), which produces directories (annually, biannually, or periodically) covering almost all major retail industries. Fairchild Publications (publisher of Supermarket News) also produces directories containing information about supermarket competition in numerous markets. There are also shopping center directories published by such organizations as the International Council of Shopping Centers, the Urban Land Institute (refer to Appendix A), and private vendors, many of which contain the listing of all retail tenants in various types of shopping centers. Such directories are often categorized by regions of the country, type and size of shopping center, and anchor tenants. Finally, "list brokers" can provide data by Standard Industrial Classification (SIC) code or by various areas of geography. This data will include "key" contacts at headquarters or store levels. Such data is generally available in a variety of digital formats for data processing or directly usable in GIS software. Other specialized list brokers can provide data on store locations and operating characteristics for supermarkets, drug chains, mass merchants, and membership retailers. Such published data must be used with caution; our experience is that such data is sometimes inaccurate both in terms of its positional accuracy or operating characteristics, and may not be up-to-date. As with all data, the analyst must validate a sufficient enough sample of the data to gain confidence in its quality and accuracy.

There are also industry-specific sources of competitive data. A relatively new source of growing importance is the Internet. Many retailers have a Home Page which often lists both the number of stores they operate and their locations. Often a telephone call to a chain's headquarters can produce a list of their retail locations.

Competitive data can also be provided by local newspaper advertising departments, municipal and county governments, and Chambers of Commerce. Newspaper ads for a chain of stores will often list all their locations in the area, as do television and radio ads. There are also magazines, newspapers and other periodicals that serve particular retail industries (such as Supermarket News, Discount Store News, Home Center News, or Restaurant Trends).

Just as it is important to gather as much information as possible about existing competition, so also is it important to learn as much as possible about expected competitive moves in the future—such as planned remodels, expansions, relocations, new stores, or even store closings. While such information is normally difficult to ferret out, municipal planning and zoning departments, zoning inspectors, building permit departments, and newspaper advertising departments often have this type of information before it becomes general knowledge. Often, a brief visit by the research analyst who asks the right questions of these sources will result in some indication as to coming changes in the competitive environment—changes such as new stores coming into the market, store remodels or expansions that are planned, store relocations that have been approved, and so on.

Site Characteristics. Site characteristics data tend to be qualitative, rather than quantitative but, they are no less important in assessing the sales potential for a retail store. Therefore, it is incumbent on a site or location research analyst to evaluate the characteristics of a site critically, and to build such observations into the analysis.

While the relationship between a store's site characteristics and its level of sales is usually not measured quantitatively (although there are instances where such measurements are gathered and used as part of an analysis, most notably, quick-service restaurants), there is little doubt that site characteristics play an important role in whether a store achieves a profitable level of sales. After all, a retail store without visibility will be difficult for prospective customers to find; a store without adequate parking (either in capacity or in configuration) will be limited in its ability to produce sales; and a store that has difficult ingress/egress will be limited in its ability to attract customers.

To the extent that a detailed market study is undertaken for a proposed retail location, it is important to evaluate the site itself relative to its characteristics. Visibility is one aspect of a site that can affect the store's performance. While it is true that consumers in an area eventually find out where stores are located through marketing, advertising, and word-of-mouth, there is little doubt that good visibility from as many directions and as great a distance as possible is a positive influence in helping customers find a store sooner. Further, visibility is particularly important in new growth areas, where the vast number of potential customers are relative newcomers to the area.

To the location analyst, visibility should be evaluated critically in terms of several characteristics. Can the site be seen from all directions? Can the site be seen from far enough away to permit safe slow-down and entry into the parking area? Are there impediments to visibility, in any direction or over distance? If so, are they natural or man-made, and can they be removed, in order to enhance visibility? Can the overall visibility be improved with signage, changes to landscaping, removal of structures, or by other measures?

Parking is another site characteristic that needs careful evaluation—in terms of capacity and configuration. There are several guidelines against which a site's parking should be measured. For example, the International Council of Shopping Centers has suggested that, for a shopping center, there should be at least 2.2 square feet of parking for every square foot of shopping center space. This translates into about 5.5 cars per 1,000 square feet of shopping center space. For supermarkets, the recommendation is that there be at least 3.0, and preferably 3.5, square feet of parking for every square foot of supermarket space (or 7.5 to 9 cars per 1,000 square feet). Further, it is generally understood that optimal supermarket parking lies within a radius of about 300 to 350 feet from the front door of the store, and within sight of it.

Ingress and egress refer to the places whereby shoppers can enter or leave a parking lot in order to shop at a particular retail store or shopping center. Ingress/egress points need to be evaluated in terms of the ease or difficulty with which customers can utilize them. In doing so, the site analyst should consider the speed limits along frontage streets, the traffic controls which govern ingress/egress points, whether or not acceleration/deceleration lanes exist, how many ingress/egress points exist, and generally how easy they are to use. For example, an ingress/egress point on a one-way street severely hampers customers coming to the store from one direction, while facilitating customers coming from the other. On the other hand, a retail site at an intersection with frontage on both cross-streets, with several ingress/egress points on each street, with left-turn and deceleration lanes on both frontage streets, and with signalized traffic controls, would be considered to have very good ingress/egress; such a site would be comparatively easy for customers to enter and exit.

Market Characteristics. Market characteristics refer to conditions that could potentially affect the well being of the retail environment. This information may be quantitative or qualitative in nature, including such topics as local economic conditions, unemployment, the composition of the employment, changes in road patterns, industrial development, growth or decline in population, and so on.

It is important to gather such information as part of a location research study, as it may affect the long-term consequences of a location decision. For example, a market in which there is a rising unemployment rate brought about by industrial stagnation may represent a good short-term retail opportunity but a weak opportunity long term. Conversely, a market that is considered marginal today might look much more viable in the future.

Information on such topics can generally be obtained from a number of sources. Many departments of municipal government deal with such data on a regular basis, including zoning, planning, economic development, and building permit departments. Regional planning commissions also maintain useful data. Other sources include colleges and universities, industrial development commissions, newspapers, and civic groups. In growing areas, considerable information regarding the anticipated direction and amount of growth can be obtained from local utilities: departments of water and sewer, electric companies, water companies, and telephone companies.

Another aspect of market conditions concerns trade area access—the ease or difficulty with which prospective shoppers can travel from their place of origin to a store or shopping center. Access is generally defined in terms of streets and highways, and the manner in which their particular characteristics add to or detract from a store's ability to draw customers over distance. It is incumbent upon the location analyst to identify and assess the major access routes that exist—freeways and surface streets, arterials, and neighborhood feeder streets.

In evaluating access, consideration should be given to barriers (natural or man-made, real or psychological), as they can cause hindrances in what may otherwise appear as unimpeded access. For example, arterial highways that are particularly busy during certain times of day can hinder their use as a means for getting to particular stores, at least during those times. Rivers and mountain ranges can limit trade area reach, while high-speed expressways can extend trade area distances.

Road characteristics (width, number of lanes, turning lanes, traffic controls, speed limits, etc.) are also important in evaluating access.

Further, access attributes are typically evaluated differently for a convenience-oriented store than for a destination-oriented store. For example, a large format destination store (such as a large home center) typically serves to a relatively large trade area, thus having the need for a strong system of freeways and major arterials that extend over great distances. A convenience-oriented retailer (such as a supermarket or local drug store), by contrast, is less concerned with marketwide accessibility, while the local system of neighborhood streets and highways is critically important. Toll roads might be considered major barriers with respect to convenience retailers, while they usually have no limiting effect on specialty and destination retailers.

Consumer Expenditure Potential. Several of the most commonly used location research sales forecasting methodologies are based on the concept of market share. That is, they require estimating the level of consumer expenditure potential that exists for the proposed store for which the sales forecast is being prepared. Such expenditure potential estimates are typically expressed as dollars per capita or dollars per household which, when multiplied by the density (people or households) of the relevant trade area, results in an estimate of the level of sales potential for the specific trade area. Sales for the proposed store are then projected on the basis of market share — the store's anticipated share of the available expenditure (sales) potential. Hence, it is important to correctly estimate the level of potential that exists, against which a market share will be applied.

There are several sources for expenditure potential information. Supermarket potential is usually expressed as PCW—per-capita weekly food store expenditure. The base data for calculating supermarket PCW in the United States comes from the U.S. Department of Labor, Bureau of Labor Statistics. The BLS conducts annual diary surveys, in which thousands of U.S. households keep diaries of every expenditure made by every member in the household. Each participating household also reports its demographic and socio-economic characteristics, such as household size, educational attainment, racial composition, income levels, occupation types, and so on. These household qualifiers are then regressed against expenditure patterns in order to relate food store type merchandise expenditures to various demographic and socio-economic characteristics. The result of this analysis is a formula that calculates PCW estimates according to household size, type of market (urban or rural), region of the country, and several other variables. Thus, when a supermarket location analyst has determined the trade area to be served by a proposed supermarket, he or she can determine the various socio-economic characteristics of the trade area, and calculate an appropriate PCW for the trade area in total, or for each of its various sectors.

However, such formulas have an inherent flaw in their application. While the formulas are calculated based on regional averages, their application is typically at a trade area scale. For example, the BLS diary surveys are a stratified sample in which the country is divided into census regions and, within each region, according to urban or nonurban categories. In other words, the analyst may be applying average expenditure data for the Midwest region for all urban areas (inner city, as well as suburban areas) to estimate the potential for a block group in a highly dense urban ethnic neighborhood. This profound shift in scale of analysis can result in startlingly inaccurate estimates.

Another source of expenditure potential in the United States is the U.S. Department of Commerce, Bureau of the Census (refer to Appendix A). This agency, in addition to the decennial Census of Population and Housing, conducts a Census of Business every five years (in years ending in 2 and 7). Included in this census are the Census of Retail Trade and the Census of Wholesale Trade which can be particularly useful to the location research analyst.

The Census of Retail Trade gathers and presents numerous types of information regarding sales, profits, employment, and operating characteristics of retail firms throughout the U.S. The tabulations of this data are presented in numerous ways: by geographic area, by type and size of retail organization, by size and type of market, by SIC (Standard Industrial Classification) code (a means of classifying retail, wholesale, manufacturing and service industries), and by merchandise line (category).

For example, using the Census of Retail Trade, the location research analyst can determine the level of snack food sales achieved by all drugstores in 1992. Or he may choose instead to look at the total sales of health and beauty aids in Metropolitan Phoenix in 1992, regardless of store type.

The major limitation of the Census of Business data is that it is only produced once every five years. Further, there tends to be a long period of time between the time when the data are gathered and the time when the results are published. While the demographic data vendors are beginning to produce intercensal estimates for some of this data, it is not with the same frequency or reliability as with the population and household data.

It is important to recognize that the amount of data needed, and its accuracy at the local level, are dependent on the type of research investigation being undertaken. A macro-level site screening study does not need data accuracy to be as great as a micro-level sales forecast study. Generally speaking, a macro-level study does not require fieldwork, relying instead on secondary data. But a micro-level study depends upon detailed data with a high degree of accuracy and reliability; fieldwork in such studies results from the necessity of gathering primary data that is up-to-date, relatively accurate, and readily available at the field level.

But whatever the purpose for the data being gathered and analyzed, and whatever the level to which the data are being used, one maxim is forever applicable: "garbage in, garbage out." To the extent that the data used in a study is sound, the results of the research investigation have the greatest chance of being reliable. But to the extent that the data are questionable in their accuracy, the results of the research may also be questionable, often leading to erroneous conclusions and inappropriate store location decisions.

3 Database Development

RETAILERS GENERALLY CONSIDER "BRICK AND MORTAR" mistakes to be among the most costly and difficult to rectify. Few decisions have such a long-term impact on a retailer's performance as a poor real estate decision. In this context, it is not surprising that many real estate decision-makers have developed site evaluation techniques to help ensure that they select locations that the company can live with for the long term.

The most basic of these approaches involves the real estate decision-maker comparing the characteristics of a site against those of existing stores with which he or she is familiar. (The thought process being used is: "The site is located next to a mall and has good access; therefore, it should do about as much as Stores A, B, and C which have the same characteristics"). It is not always reliable. Some decision-makers may attempt to refine this process by incorporating demographics into the site evaluation process, usually by examining a demographic profile for an arbitrary area (e.g., a five-mile radius) around the site, and comparing it with the comparable data for existing stores.

When using these approaches, decision makers are unwittingly employing a crude store database to assess the viability of new site opportunities. While they are on the right track, their approaches fall short, as they do not address many of the basic issues that are known to influence store performance. For example, this approach does not consider how far consumers are willing to drive to shop; a five-mile trade area was assumed. Research has consistently demonstrated that trade areas are not conveniently broken into concentric circles; rather, consumers will drive great distances when they have no logical shopping opportunities, and conversely, short distances when more convenient competitive alternatives are present. Further, one must consider how consumers are distributed throughout a trade area. Again, research has shown that stores with a proximate consumer base generally experience stronger sales performance relative to stores whose consumer base is concentrated at the periphery of their trade areas.

While using a crude database as a benchmark against which to evaluate future site opportunities is better than a "roll of the dice," its reliability is limited because of assumptions that are made regarding critical factors such as trade area size, customer profile, the relationship between consumers' propensity to make purchases and the distance they must travel, competition, and so on. It is also unnecessary to make such assumptions. Today, using Point of Sale (POS) information, demographic databases, and other readily available sources and services, the influence of factors such as demographics, accessibility, and competition on sales performance can be easily quantified.

Building a Store Database

By quantifying these issues and incorporating them into a store database, powerful site evaluation tools can be developed. The decision-maker now has the ability to determine, for example, whether the strongly favorable demographic makeup of a site's trade area will more than compensate for the intense competition the site will encounter. These sophisticated store databases are the foundation from which successful site evaluation systems are developed.

A store database is usually comprised of information specific to each store, which typically includes (but is not limited to) customer and transaction data, competition, the size of each store's potential customer base (e.g., population, households), demographic characteristics, and distance measurements between trade area sectors (e.g., census tracts, ZIP Codes) and the database store. The process by which this information is collected can be a formidable obstacle to the development of a store database. The processes used to collect this information, the methodologies commonly used to obtain it, and the various sources for data that are typical components of a store database are addressed in the first half of this chapter.

The second half of this chapter addresses the logistics associated with assembling a store database. Once it has been assembled, the relationship between the database components and store performance can be measured, thereby allowing the real estate decision-maker to focus on the variables and characteristics that have consistently been found to be positively associated with store performance. While the specific components of the database will be different for every retail concept, the database development process is generally consistent throughout a broad retail spectrum.

The Data Collection Process

The information that is used in developing a database (and from what source it is derived) is a function of how quickly a decision needs to be made, the level of accuracy required, as well as developmental cost considerations. As an example, a retailer facing the pressure of short-term property options may not have the time necessary to acquire and apply all of the data that should be included in a comprehensive analysis, and instead may opt for a cruder and more cursory analysis. While simplistic, these approaches may be adequate when a quick decision is required and no database is available; however, they are not endorsed as bona fide procedures for evaluating existing and proposed site opportunities as the long-term financial ramifications of most site decisions are too extreme to be relegated to such cursory analysis. Rather, a more rigorous analysis using a database of existing stores for which the relationships between sales performance, demographics, and competition have been statistically measured, represents the preferred approach when a consistently accurate assessment of a retail sites' viability is required.

Most retail store databases are comprised of several datasets which typically include population/household data and basic demographic information, as well as information pertaining to the competitive environment, access to the site, site characteristics, market or regional characteristics, associated with each database store. Much of this information (particularly population and demographic data) can be purchased from government and private sources, and because of this, the challenge is not in obtaining data; rather it is in knowing which sources can provide the data which is best suited to address each retailer's unique requirements. Additionally, factors such as how frequently data is updated, the accuracy of the population and demographic forecasts, and the source of the base data are important considerations when evaluating data vendors. It is important to remember that when using demographic information obtained from vendors, usually only the population, household count, and income data (and in some instances, age and ethnicity data) are updated to current and future levels. Most other demographic information (occupation, education, housing characteristics, etc.) reflects the most recent Census. Therefore, an area profile summary report, which indicates that 60 percent of all households are owner-occupied in reality, reflects the proportion of owner-occupied households present in the area during the most recent Census. Updates for selected demographic information may sometimes be available from local sources, although there is no consistency regarding the types of variables updated.

If updated demographic information is not required, it may not be necessary to purchase data from a vendor at all. Census information is available without charge; the only effort required to retrieve it is a trip to a Census repository library or to access it via the Internet. Conversely, for retailers operating in very rapidly growing areas, data available from the sources cited above may not be appropriate, as the accuracy of their population and housing data in high-growth areas is less than accuracy levels for more stable areas. This is because data vendors cannot practically "keep up" with every emerging high-growth area in every market. For example, population information obtained from data vendors for exurban areas which historically were rural or agricultural, will likely reflect historical growth trends, even though, subsequent to the completion of the most recent census, they have emerged as major areas of suburban

FIGURE 3-1
DEMOGRAPHIC VARIABLES

Population — Census Year, Current Year, and Projected Year

Group Quarters Population — Residence of 10 or more unrelated individuals such as college dormitories military barracks or institutions; Census Year, Current Year, and Projected Year

Households — Census Year, Current Year, and Projected Year

Race — Reflects the self identification by Census respondents rather than a scientific definition of race; mixed race categories to be added to the 2000 Census; Census Year, Current Year, and Projected Year

Hispanic Origin — Like race, a self-declared variable; persons of Hispanic origin may be of any race; Census Year, Current Year, and Projected Year

Age — Category by Sex; Census Year, Current Year, and Projected Year

Housing Unit — Single and Multiple Number Unit designations by ownership status; Census Year

Income — Households by Income Category, median and average income; Census Year, Current Year, and Projected Year

Family Status — Married, Single Householder and Head of Household (Male or Female); Census Year

Females with Children — Females aged 16 or older by children age group (0–5, 6–17); Census Year

Educational Attainment — For all people aged 25 or older by highest level; Census Year

Occupation — Employment Categories for persons 16 years and older and white or blue collar categories; Census Year

development. In these circumstances, retailers are advised to seek population and housing data from local sources (local planning departments, councils of government, building departments). Generally, these departments have up-to-date information on current and anticipated population and household levels at a subgeography level (e.g., census tracts, ZIP Codes, traffic zones, etc.). Figure 3-1 summarizes the most common types of demographic variables that are considered in the development of a store database.

Competition is a factor which significantly influences the site selection process, and is also an important component of databases which are used to evaluate future locations. In incorporating a competitive dataset into a store database, one must first decide which stores (or types of stores) are, in fact, "competitive." For example, are service stations which offer convenience groceries such as bread and milk "competitive" with contemporary supermarkets? Are drugstores which offer convenience hardware and hand tools "competitive" with warehouse home improvement stores? There are logical and practical considerations with respect to which stores or types of stores should be included in a competitive dataset. No retailer has the resources needed to inventory and assess every store that has merchandise overlap with their concept. Thus, in deciding which competitors to include in the dataset, retailers should focus on those stores which have a marked impact on the performance of their stores ("direct" competitors); the remaining competitors should be considered indirect, and addressed as competitive "noise." As a general rule, the level of indirect competitive noise is consistent from market to market; because its impact on store performance is usually negligible and consistent, competitive noise does not have to be directly addressed when evaluating the viability of future site opportunities.

General information (location and size) regarding the most common types of retailers (supermarkets, department stores, quick service restaurants) can be purchased from numerous data vendors; however, vendor data usually does not include the level of detail desirable in a comprehensive competitive dataset (hours of operation, merchandising strengths, cotenancies, operational strengths, visibility). Further, competitive information obtained from vendors (the location and sizes of competitors) usually needs to be verified and updated to reflect recent changes. Therefore, it is usually necessary to conduct primary research in order to ensure that competitive data for each database store is as up-to-date and complete as possible. Real estate management or brokers familiar with the database store markets are usually reliable sources for verifying existing competition.

At the risk of stating the obvious, the types of information that should be included in a competitive dataset should be a reflection of the factors which make each respective competitor "competitive." For example, for certain types of retailers (e.g., athletic footwear stores), the types of brands offered by competitors directly influences the degree to which they are competitive. Alternatively, for retailers that primarily sell commodities, price tends to be the factor that most strongly influences the degree to which they are competitive. Other factors which are commonly included in competitive datasets include location, location type (freestanding, shopping center, mall), size, merchandising strengths, hours of operation, operational strengths, estimated volume, sales, and so on. It is obviously impractical to list all of the factors which may be appropriate components of a competitive dataset as they are as varied as is retailing.

Figure 2-1 in Chapter 2 provides an example of a typical competitive information card used to collect pertinent competitive data. As mentioned previously, much of this information is <u>not</u> readily available through secondary sources. Thus, if a complete competitive dataset is required, much of this information must be collected via primary research—that is, fieldwork, which typically involves visiting each competitor and assembling appropriate information. Typically these visits include price checks, an assessment of merchandising/operations strengths, and estimates of store size and sales.

Ultimately, the location of competitors encountered by each database store should be identified on a map (refer to Chapter 5, Map 5-1). Further, the information unique to each competitor should be retained (either hard copy archival or in a computer-based database retrieval system) so that its impact on store performance and, ultimately, the performance of future locations can be assessed (an elaboration of competitive issues, as well as specific methodologies used to assess competitive impacts on store performance are addressed in Chapter 5).

Datasets which relate to customer information (how much they spend, where they reside, how far they drive to shop each database store, e.g.), are essential components of a site evaluation database. For larger retailers, customer information represents a huge volume of information which often necessitates a large database structure, such as that provided by a data warehouse. In contrast to demographic information which is readily available from vendors, data specific to a retailer's customers is usually obtained from customer exit surveys or point of sale (POS) information (consisting of customer addresses and transaction sizes collected during the checkout process), or from information obtained from credit card or check purchases, or in-store "preferred customer" programs. The accuracy of this information is critical as it is the basis for determining the geographic extent of each store's trade area; and in estimating how each store's sales are distributed throughout its trade area (these issues are discussed in detail later in this chapter).

Because credit cards and check data represent an inexpensive and relatively easy means of collecting customer information, retailers are often enticed to use these records as the basis for defining store trade areas and determining the sales distribution of each store. However, research suggests that using credit card purchases as the basis for collecting customer information often results in an under-representation of lower income-customers. Credit card and check transactions tend to be associated with large purchases, which, in turn, are likely to reflect a more extensive trade area. Therefore, retailers who are considering these approaches should first determine if data collected from credit records are truly representative of their actual customer base. To do this, a representative sample of stores should be selected. This sample allows the retailer to compare the customer and sales distributions implied from customer exit surveys or POS information (which reflects cash as well as credit purchases), against the comparable data using credit records exclusively. If the customer and sales distributions are alike, then credit card and/or check information would probably be acceptable for the purposes of defining trade areas and estimating sales distributions throughout each trade area.

Another (although less accurate) method of collecting customer information is license plate surveys. This approach is particularly useful in helping a retailer understand the trade area served by a competitor, or for retailers who are modeling their business after an existing chain. In either instance, the process simply involves collecting the license plate numbers of vehicles parked in the lots of competitors. For a modest fee (sometimes as little as a few dollars per license plate), many states will provide vehicle owner license registration addresses from which the store's trade area can be implied.

However, there are pitfalls to this approach. The information procured pertains to the addresses at which the vehicles are registered; in the case of a company car, this address likely is the driver's workplace. Further, vehicles are only registered annually; in high-growth areas, many of the addresses collected could reflect the driver's former place of residence. Finally, as a transaction size cannot be linked to the customer, every license plate address must be considered of equal significance, which may result in misdefined trade areas if there are both convenience and shopping goods components to the trade area. Because of the shortcomings associated with using vehicle registration addresses as the basis for customer information, a license plate survey should only be employed if other alternatives are not available.

To this point, our discussion concerning the collection of customer information has focused on linking each customer to their place of residence. This is because for most retailers, consumer shopping behavior is directly related to the proximity of the consumer's place of residence to each store. However, some types of retailers are less dependent on the residential base and more dependent on the "daytime population" which surrounds their stores, or "transient population" (the number of vehicles which drive past their stores daily). As an example, quick service restaurants often depend on nearby generators of lunch-time crowds, such as office parks and other commercial facilities for a significant proportion of their business. Similarly, office supply companies are dependent on surrounding office (rather than residential) developments as sources of sales potential. In such instances, information regarding the origin and destination of the customer's shopping trip is more useful than the customer's place of residence. As with "residential-oriented" retailers, retailers dependent on daytime population can use information pertaining to the daytime location of customers as the basis for determining the geographic extent of each store's trade area, and in measuring how each store's sales are distributed throughout its trade area.

The sources for demographic and competitive data, and the methodologies used to collect customer information discussed above are not intended to be all inclusive, nor are they appropriate for all retailers; it would be impractical and inappropriate to compile sources for virtually every category of retailer in this publication. As an example, one of our clients is a boating equipment and supplies retailer. Information regarding its customer base, boat owners, is not available through traditional sources. As a result, alternative data sources were explored until ultimately, one suitable to the client's needs was discovered. Thus, we have intentionally focused this narrative on the information and data that is commonly included in store databases, and sources and methodologies commonly used to obtain such data. Retailers who have an unusual customer appeal or serve a very narrow segment of the retail sector are strongly encouraged to research government and industry sources for consumer information unique to their retail niche.

Assembling a Store Database

Up to this point, we have focused on what and how to obtain the information used in building a store database. The balance of this chapter addresses the process by which the database is built.

Generally, the first step in developing a database is to identify the stores which represent the best database store candidates. There are numerous practical as well as methodological issues that should be considered when selecting candidates for inclusion into a store database. First, there is a strong relationship between the time and expense associated with database development and the size of the database. Therefore, it is recommended that a database development budget be established, and that the budget be considered when determining how many stores should ultimately be included in the database.

Second, the purpose of developing the database should be considered when selecting stores for database inclusion. In this context, it is important that the stores included in the database reflect the types of stores and types of locations which will be assessed in the future to the extent that it is practically possible. More specifically, if all future stores will be 15,000 square feet, avoid including 7,000-square-foot stores in the database. If all future stores will be developed in the Northeastern United States, avoid a database comprised entirely of stores from other parts of the country; economic considerations (unemployment rates, the nature and well-being of industry/major employers, etc.) unique to specific markets or regions of the country can significantly influence retail performance.

The site and situational characteristics associated with database store candidates should also be considered as these can have a significant influence on the performance of existing and future stores. These factors are particularly important for convenience-based retailers (quick-service restaurants, convenience stores, drug stores, supermarkets, etc.). The reason for this is that among convenience retailers, consumers usually choose to shop the store that is closest, easiest to shop, and has the lowest prices. As such, a store that has easy ingress/egress, is highly visible, has ample parking immediately proximate to the store entrance, and so on, has a distinct advantage over other convenience retailers with inferior site characteristics. For destination-oriented retailers, such as a high-end apparel or jewelry store, factors such as visibility and ingress/egress have much less of an impact on consumer shopping habits and should be looked at as limitations on, rather than determinations of, sales potential.

The types of issues that are typically considered and noted when selecting stores for inclusion in a database may include drive-through windows, the number of ingress/egress points, visibility, the adequacy and proximity of parking, adjacent retail and commercial developments, and the nature of the site (freestanding, in-line, in mall). Further, the accessibility patterns associated with each store should be considered as they have implications with respect to the geographic extent of the trade area served by each store. More specifically, a database comprised exclusively of stores on two-lane roads in neighborhood locations, would be of little value in analyzing the viability of sites on major commercial arterials or near freeway interchanges (and vice versa). Thus, the nature of the road network serving each database store candidate should be considered (the number of lanes, traffic controls, signals, turn lanes, speed limits), as these factors impact the drawing power of each store. Impediments to access such as road medians, one-way streets, physical and cultural barriers (rivers, differing ethnic/cultural neighborhoods, industrial areas, etc.), and even excessive traffic congestion should also be considered when selecting database stores.

As information regarding the site and situational characteristics of each database store cannot be obtained from a consistent and reliable source, it is usually necessary to conduct primary research to collect site and situational information; that is, each store is usually visited for the purpose of collecting this information firsthand. Often, this primary research is conducted concurrent with the collection of competitive information for each database store (this process was discussed earlier in this chapter). By including site, situational, and accessibility variables in the database, the research analyst will ultimately be able to quantify the degree to which they influence store performance (refer to Chapter 5 for more details).

In selecting stores for inclusion in a database, it is strongly recommended that retailers resist the temptation to "stack the database deck" with the strongest performing stores. It is just as important (perhaps even more important) to analyze and identify the factors that limit sales performance as it is to understand which factors are associated with strong performance. Similarly, stores whose sales performance is influenced by unusual circumstances (they have the best/worst manager in the chain, their hours of operation are limited due to lease restrictions or blue laws, they are receiving an inordinate amount of marketing support) should be avoided.

The "maturity" of the stores which comprise the database should also be considered. In other words, the research analyst should avoid developing a database comprised of very new, as well as more mature stores. This is because most retailers undergo a period of maturation (typically two to five years) during which significant changes can occur with respect to store performance and trade area extent. Since a database is used to quantify relationships that exist between store performance and factors such as the distance consumers travel to shop, demographic characteristics, and competition, the database should ideally be comprised of mature stores only. In instances, where this is not practical (e.g., a very young retail concept), the analyst should strive to ensure that the age of the database stores is roughly comparable, thereby ensuring that any biases resulting from their immaturity are at least consistent from store to store. Ultimately, the goal of database store selection should be to ensure that the database is representative of the types of stores and types of locations that will be developed in the future. There is no such thing as a database that is "too small." Much can be learned about the factors which influence store performance from even a one or two store database; obviously however, within the practical considerations discussed above, more is better.

Having selected a database comprised of appropriate stores, the next step in developing a database is to process the customer and transaction data collected using POS surveys (or one of the other methods described previously). The purpose of processing this data is to determine how a store's sales are distributed geographically, so that a trade area for each database store may be defined. Further, this step facilitates an analysis of the relationships that exist between store performance and the geographic distribution of key factors (population, demographics, competition, etc.) that may ultimately be conducted. It is important to note that only customers who have made a purchase should be considered in this analysis. Browsers are not necessarily customers, and their inclusion in customer data could exacerbate efforts to identify relationships that exist between sales and consumer demographics.

Typically, a store's sales are distributed according to some convenient unit of geography such as ZIP Codes, census tracts, or block groups (these represent common census and postal geographies in the United States, and are referred to hereafter for illustrative purposes). The most appropriate unit of trade area geography for a particular retailer is a function of the size of the trade area it serves. Stores with large (roughly eight miles or greater) trade areas (such as destination-oriented category killers), could have hundreds of census tracts within each trade area, rendering data analysis and interpretation unwieldy. Further, in such circumstances, the potential for sample error for individual geographic components increases as the requisite customer sample size is directly related to the number of pieces of geography comprising a trade area. For stores serving large trade areas, ZIP Codes are usually the most appropriate unit of geography.

Conversely, convenience-oriented stores (such as supermarkets and drugstores) usually serve trade areas that extend no more than four miles, and often are two miles or smaller. For such stores, census tracts or block groups are usually a more appropriate basis for distributing store sales.

There are strengths and weaknesses associated with the various units of geography which are used to comprise the trade areas of database stores (e.g., ZIP Codes, census tracts, and block groups). Most of these units of geography can be readily associated with the substantial demographic databases available from vendors or the U.S. Census. However, as stated in Chapter 4, a specific advantage of ZIP Codes as it relates to acquiring customer data at the point of sale, is that customers can readily provide this information during the check-out process; few customers know the census tract in which they reside. Customer addresses can be "geocoded" to their corresponding census tracts (using any number of geocoding services provided by data vendors and other sources), but this a more expensive and time-consuming alternative to collecting ZIP Codes. It is much easier to collect a ZIP Code than an address during the check-out process. Moreover, consumers are generally more willing to provide retailers with their ZIP Code than their exact address.

A "downside" to using ZIP Codes to define trade areas for database stores is that approximately 10 percent or about 4,000 of the roughly 43,000 ZIP Codes in the United States change each year. These changes can involve boundary shifts, assignments of new code numbers to an area, or subdivisions of former ZIP Codes. Keeping up with this voluminous change is part of the reason that virtually no vendor will guarantee that its data is fully in accord with current ZIP Codes. Further, some ZIP Codes collected from customers are useless because they represent post office box locations. In these instances there may be no information indicating whether or not the ZIP Code given pertains to a residential or business post office box. There are ways to get around these obstacles, such as directly calling post offices to obtain the most recent ZIP Code boundaries. Nonetheless, these problems are an impediment to an otherwise very convenient means of collecting customer information.

As discussed above, the major advantage of using census tracts to define trade areas for database stores is that they are relatively small, and therefore, are convenient units of subgeography for retailers who serve small trade areas. However, to group sales and customer distributions by census tract, the customer address must be obtained and geocoded; as previously mentioned, some consumers are uncomfortable providing such information, and further, relative to collecting ZIP Codes, this is a time consuming and expensive endeavor. As with ZIP Codes, census tract boundaries can also change over the years. On occasion, the definition and boundaries of an entire metropolitan area may change, as in the early 1980s when Monroe County, Michigan was eliminated from the Toledo, Ohio Metropolitan Area and placed into the Detroit, Michigan Metropolitan Area.

Whatever unit of geography is used to organize data (ZIP Codes, census tracts, block groups, etc.), all necessitate the assumption that the arithmetic means, medians, and modes which are used to characterize the demographic composition of each unit of geography, reflect the demographics specific to customers originating from each unit of geography. The implication of this to forecasting is that the more important a specific demographic segment is to forecasting, the more important the internal consistency of the demographics within the geographic unit becomes.

As discussed earlier, the most appropriate unit of trade area subgeography (e.g., ZIP codes, census tracts, block groups) is a function of how large (geographically) the database store's trade areas are. Thus, retailers must anticipate the typical size of the database stores' trade areas in order to determine whether to collect customer and transaction information at the ZIP Code level, or whether this information must be generated for a smaller level of trade area subgeography. Once an appropriate unit of subgeography has been identified, the process of defining trade areas for each database store may begin.

If a retailer does not have a "feel" for the geographic extent of its trade areas, then an analysis of a few (four or five) database stores should be conducted in order to better anticipate trade area size. The "test" stores should reflect locations that are typical of the stores, which will ultimately comprise the complete database (e.g., suburban versus urban, freestanding versus in-line, in-mall) so that the impact that access, population and cotenancies have on trade area size is reflected. The trade area for each of these database test stores should be defined using a simple ZIP Code analysis as summarized below.

To estimate a store's trade area extent using ZIP Codes, customer ZIP Code and transaction data should be collected in each test database store using one of the approaches discussed earlier in this chapter (e.g., POS surveys, exit surveys, credit records, internal database). Ideally, every customer should be surveyed for a one year period to provide a "perfect" sample. While this is an unrealistic goal for most retailers, it is important to remember that the closer to a 100 percent sample a survey is, the more accurate the conclusions derived from it will be.

More practically, these surveys should be conducted with a <u>minimum</u> of 400 to 600 customers (but preferably 1,000 customers or more) as they are exiting the stores. The surveys should be conducted for at least one week (longer is better), and distributed proportionate to the stores' sales distribution. That is, if 20 percent of a store's sales are generated on Fridays, then 20 percent of the surveys should be conducted on Friday; if 35 percent of a store's daily sales occur during morning hours, then 35 percent of the daily interviews should be conducted during the morning. Interviewing during atypical situations such as an intense marketing campaign or holidays should be avoided. Simply stated, the goal of this process should be to collect as many surveys as is practically possible, while concurrently ensuring that they are indicative of a typical operating period.

Ultimately, the customer ZIP Code and transaction data is used to provide an indication of how each store's sales are distributed by ZIP Code. More specifically, it is assumed that the sales distribution implicit from the customer surveys is indicative of the actual annual sales distribution for each test database store. As indicated in Figure 3-2, the first step in this process is to sum all of the sales that originated from each ZIP Code during the survey period in order to calculate the total dollar volume per ZIP Code for the survey period for each test database store. For example, if a total of three transactions (a $20, a $30, and a $40 transaction) originated from ZIP Code 12345 for test database Store #1, then a total of $90 originated from this ZIP Code during the survey period. Assuming the total dollar value of all transactions accounted for during the survey process for test database Store #1 equaled $2,000, then 4.5 percent ($90 divided by $2,000) of this store's sales originated from ZIP Code 12345.

FIGURE 3-2
SALES DISTRIBUTION BY ZIP CODE

Survey Sales from ZIP Code 12345	Total Sales from Survey	Capture Rate
$90	$2,000	4.5%

Thus, by dividing each ZIP Code's sales by the total sales collected throughout the survey period, the proportion of each test database store's total volume that originates from surrounding ZIP Codes can be estimated. This proportion is commonly referred to as a "capture rate" as it represents an estimation of the proportion of store sales that are "captured" from surrounding ZIP Codes.

Once the capture rates for ZIP Codes which surround each of the test database stores have been calculated, the geographic distribution of each store's capture rates should be examined to determine whether the majority of each store's sales originate from only one or two ZIP Codes. If this is the case, then ZIP Codes are an inappropriate unit of geography from which to build trade areas for the remaining database stores, because the database will ultimately be used to quantify the relationships that exist between store performance, and distance, demographics, competition and other pertinent factors (e.g., which income levels are associated with strong store performance, how does distance impact store performance levels, which competitors most severely inhibit store performance). If the trade areas of database stores are comprised of fewer than, for example, six pieces of subgeography, there are usually not enough different examples to allow these relationships to be statistically established (a discussion of the techniques used to statistically quantify these relationships is presented in Chapter 4). Further, in application, such a forecasting system would not provide for much forecast stability, as the capture rates would not reflect an accurate distribution of sales. Also, from a more practical perspective, the number of database observations from which a forecast could be based would be limited.

Supermarkets (and most other convenience-oriented retailers) are typical of stores whose sales primarily originate from only one or two ZIP Codes. In past research, we have identified many instances where a single ZIP Code accounted for 60 percent of a supermarket's volume; in effect, the trade areas of these stores consist of a single ZIP Code. In such circumstances, a unit of geography smaller than ZIP Codes should be used for the purposes of defining trade area extent, and for the subsequent analysis that is often conducted for the purpose of quantifying the relationships between store performance, and distance, demographics, and competition. As previously mentioned, many vendors offer geocoding services which "tag" customer addresses to a specific census tract or block group. Thus, for stores with small trade areas, an additional step must be added to the customer information processing procedure; the customer addresses and the corresponding sales information collected during POS or customer exit surveys in each database store, must be first geocoded (by census tract, block group, or other appropriate unit of geography). Subsequent to this process, capture rates for each census tract (or block group) are calculated using the previously described methodology.

After capture rates for each unit of subgeography for each database store are calculated, trade areas for the database stores should be defined; typically, a map for each database store is produced which depicts the database store location in relation to "sister" stores and competition. Overlaid on this map are the geographic boundaries for the unit of subgeography that is most appropriate for the type of stores being evaluated (for the purposes of this example, we will use ZIP Codes). Maps of ZIP Code boundaries (or other geographic boundaries) can be obtained from vendors or government sources for a nominal fee. Geographic information systems are particularly useful for trade area definition and can be readily customized to automate these tasks.

To define a trade area for each database store, its capture rates (by ZIP Code, census tract, etc.) should be plotted on the database store map (Figure 3-3). A store's trade area is generally considered to be a <u>contiguous</u> area accounting for between 65 percent to 75 percent of a store's sales. This represents a <u>general</u> range; these proportions may be somewhat higher or lower depending on the nature of each specific retailer. In this context, defining a database store's trade area involves aggregating ZIP Codes, beginning with the store's "home" ZIP Code and progressing logically outward, until approximately 65 percent to 75 percent of the store's sales have been accounted for. It is usually helpful to begin this process by highlighting all ZIP Codes that account for a minimal proportion of store sales (e.g., 1 percent or more). Examining this area and aggregating the capture rates for the contiguous area around the

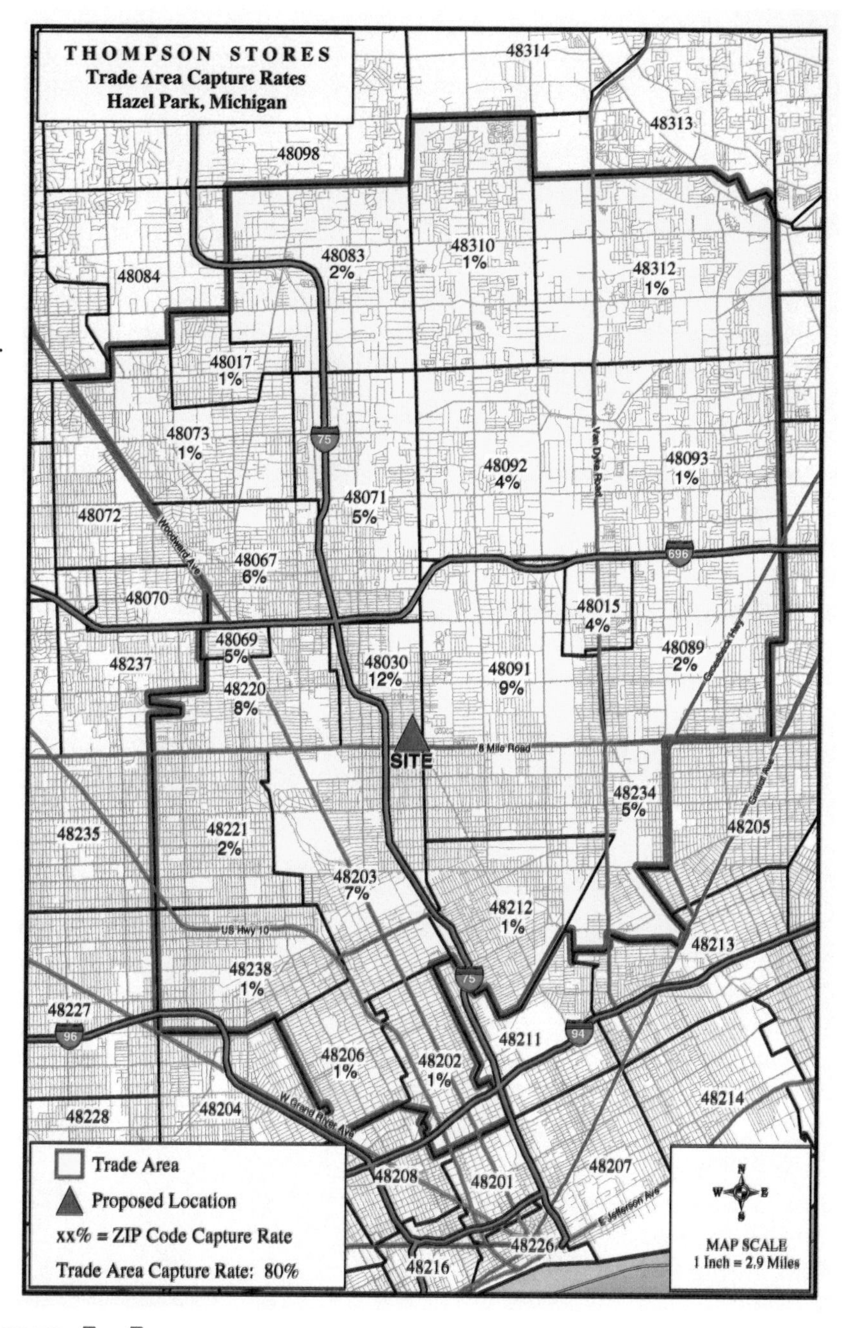
THOMPSON STORES
Trade Area Capture Rates
Hazel Park, Michigan
48314
48313
48098
48083
2%
48310
1%
48312
1%
48084
48017
1%
48073
1%
48092
4%
48093
1%
48071
5%
48072
48067
6%
48070
48015
4%
48069
5%
48237
48030
12%
48091
9%
48089
2%
48220
8%
SITE
8 Mile Road
48234
5%
48235
48221
2%
48205
48203
7%
48212
1%
US Hwy 10
48213
48238
1%
48227
48211
48206
1%
48202
1%
48214
48228
48204
W Grand River Ave
48208
48201
48207
E Jefferson Ave
48226
48216
Trade Area
Proposed Location
xx% = ZIP Code Capture Rate
Trade Area Capture Rate: 80%
MAP SCALE
1 Inch = 2.9 Miles

map 3—3

store, will provide a quick indication as to whether the trade area is overdefined, underdefined, or appropriate. It is worth noting that trade areas will likely not be represented by neat or concentric areas of geography; rather, trade area boundaries will be irregular, reflecting factors which are known to influence and affect shopping habits such as competition, accessibility, physical and cultural barriers, and the like.

Sales not captured by a defined trade area (commonly referred to as sales from beyond the trade area or, more simply, "beyond" sales) typically originate from consumers driving unusually long distances to shop. Further, these sales also may originate from consumers who shop a particular store not because it is convenient to their place of residence but rather, because it is convenient to their workplace, church, a relative, or other reference point. While theoretically, it would be handy if each database store's trade area captured 100 percent of its sales, in practice this represents an unrealistic goal. Our experience has been that for most trade areas, a 65 percent to 80 percent capture rate usually corresponds with an efficient and logical trade area geography, beyond which a point of diminishing returns is reached; to increase the trade area capture rate from, for example, 75 percent to 85 percent, usually necessitates including ZIP Codes with insignificant capture rates (i.e., less than 1 percent). The result is inclusion of large areas of geography and population density to the defined trade area. As the primary reason for developing a database is to facilitate the analysis of relationships that exist between customer demographics and sales, it is counterproductive to include within a store's trade area, large areas of geography and large numbers of consumers that contribute little (if at all) to its sales performance. Chapters 8 and 9, which addresses the various site evaluation techniques used by Thompson Associates explain how "beyond" sales should be handled when using databases to assess the viability of future site opportunities.

Once trade areas for the database stores have been defined, the process of approximating the distance that consumers residing in each trade area segment must drive to shop the corresponding database store, begins. It has been consistently demonstrated that there is a strong relationship between the distance a consumer must travel to a store and the amount the consumer spends at the store annually. In order to quantify this relationship for the purpose of using the database in application as a forecasting tool, it is necessary to derive an accurate estimate of the distance consumers residing in each trade area segment must travel to shop the corresponding database store.

The first step in this process is to approximate the centroid (center) of the population density for each trade area segment. This process can be accomplished by simply "eyeballing" the center of the population density, or can be more precisely determined using the distribution of population within each segment at the census block or block group level.

The next step is to estimate the distance from the centroid of each segment to the corresponding database store. The simplest (but usually least accurate) means for measuring distance, is to use straight-line measurements from each centroid to the store. That is, no regard is given to the route consumers would likely take in driving to the store; these measurements are "as the crow flies." The obvious downside to this approach is that consumers usually cannot drive in a straight line from their home to a store; as such, the relationships derived between distance and sales performance based on straight-line distance measurements will likely be inaccurate.

A more accurate means of measuring distance is to estimate the driving distance from the centroid of each segment to the corresponding database store. Using this approach, the distance from the centroid is measured by following the route that consumers residing in each segment would likely take when shopping the store. As such, if a river blocks access from a particular segment, the distance from that segment would reflect whatever route that would be necessary in order for consumers to access the store.

Ideally, drive time (rather than distance), represents the best means of measuring the difficulty each segment's consumers would encounter in accessing a store; such a measurement would incorporate speed limits and traffic controls, as well as the most efficient driving route. While this method is impractical if applied manually, technological breakthroughs such as Geographic Information Systems (GIS) will make this a more practical alternative in the future. While the software to conduct this analysis is available now, very little detailed data at a trade area scale is currently available with which to perform the calculations.

Whatever approach is used, the database development analyst will ultimately use the distance measurements to quantify the degree to which increased distance inhibits the consumers propensity to make expenditures at the database stores. This information is obviously very valuable in forecasting for new locations as it allows the distribution of consumers within a proposed site's trade area to be accurately reflected in a sales forecast. Further, understanding the relationships that exist between distance and sales allows the analyst to "factor-out" the influence of distance on sales performance, and more clearly focus the remaining relationships (such as the relationships between sales and demographics or competition).

As mentioned previously, one of the primary reasons for developing a store database is to use it to forecast sales for new locations. The chapters which follow discuss the various forecasting tools that can be developed as byproducts of a store database, as well as their application. A step toward the development of these forecasting tools is the calculation of "penetration levels" or "market shares" for each unit of trade area subgeography, as well as for the trade area as a whole. Simply stated, penetration levels and market shares measure how effectively a store is serving the various geographic components of its trade area (as well as the trade area as a whole). To better understand the concept of penetration levels and market shares, the following example is provided. Consider hypothetical Store A which achieves annual sales of $10.0 million. Using customer and transaction data, it has been determined that Store A achieves a 5 percent capture rate (or $500,000 in sales annually) from ZIP Code 77777 and an 8 percent capture rate (or $800,000 in annual sales) from ZIP Code 88888. Initially, it would appear that Store A more effectively serves consumers in ZIP Code 88888 relative to ZIP Code 77777. However, suppose ZIP Code 88888 has a population of 10,000 persons, whereas ZIP Code

77777 has a population of 5,000 persons. On a per capita basis, Store A generates sales of $100 from ZIP Code 77777 whereas it achieves sales of only $80 from ZIP Code 88888. Thus, despite its stronger overall performance in ZIP Code 88888, Store A more efficiently serves consumers residing in ZIP Code 77777 by virtue of its greater per capita sales from this ZIP (Table 3-4). By expressing sales as a ratio of density (e.g., sales per capita, sales per household), a "penetration level" is calculated. As we will see in Chapter 8, penetration levels represent a very powerful forecasting tool.

TABLE 3-4
SALES PENETRATION BY ZIP CODE

Zip Code	Population	Sales	Capture Rate	Per Capita Sales
77777	5,000	$500,000	5%	$100
88888	10,000	$800,000	8%	$80

Market share is comparable to penetration levels except that the sales achieved in each unit of subgeography are expressed as a percent of the total expenditure potential present in each unit (see Chapter 9 for an explanation of expenditure potential). Using the previous hypothetical Store A example, suppose that the expenditure potential in ZIP Code 88888 was $5.0 million while the potential in ZIP Code 77777 was $2.5 million. Store A's market share in ZIP Code 77777 would be 20 percent ($500,000/$2,500,000), versus only 16 percent ($800,000/$5,000,000) in ZIP Code 88888. Thus, even though Store A generates more sales from ZIP Code 88888, it captures a greater share of the potential in ZIP Code 77777. Among retail concepts for which expenditure potential is readily quantifiable, market share can also be a powerful forecasting tool.

At this point, it is worth mentioning that store databases represent a "snapshot" of store performance for a given point in time. Therefore, databases need to be updated periodically to ensure that they accurately represent the factors that influence current store performance. It is impossible to generalize about how long a database will remain "fresh." This will depend largely on the dynamics of each segment of the retail industry. For example, if a retailer has an evolving concept (evolving store sizes, merchandise lines, pricing strategy, etc.), then the database will need to be updated frequently in order for it to keep up with this evolutionary process. The same holds true if competitors are undergoing an evolutionary process as this will affect the degree to which they are competitive. Further, changes in the marketplace (changing demographics, road networks, retail developments, population growth, etc.) necessitate database updates. In short, when the conditions implicit in a database do not reflect reality, it is time to update the database.

4 Customer Analyses

ANALYZING AND QUANTIFYING THE DEMOGRAPHIC COMPosition and shopping habits associated with a retailer's customers provide direction to decision-makers involved with all aspects of a retailer's business. For example, retail executives charged with the responsibility for making store deployment decisions can use demographic information to help decide where to locate stores within new or existing markets. There is a wealth of customer information that can be derived using some of the basic approaches described in this chapter.

There are two fundamental concepts that will be discussed in this chapter: "customer demographics" and "target customer profile." While there may not appear to be a great deal of difference between these two concepts, the distinction is actually an important one which has significant implications for future decisions concerning store deployment opportunities and marketing efforts.

The "customer demographics" for a retailer are the demographic characteristics of those customers currently shopping at the retailer's store or stores. A retailer can use a variety of sampling techniques to survey a representative cross-section of customers, ask them for selected demographic characteristics, and then aggregate that information to identify the retailer's "customer demographics." This information is valuable as it defines the demographic composition of the retailer's current customer base—as such, any merchandising or marketing efforts designed to retain sales should be targeted at this base.

Unfortunately, the "customer demographics" are often influenced by a retailer's location, and may reflect the demographic composition of the surrounding populace rather than the optimal, or ideal customer for that retailer. The "target customer profile" represents those customers who, all other things being equal, spend the most on a per-capita or per-household basis at the retailer in question. As an example, older consumers typically represent the "target customer profile" for drug store prescription counters, since older persons spend, on average, considerably more on prescription drugs than their younger counterparts.

To illustrate problems inherent with using "customer data" exclusively when making real estate and marketing decisions, the following example is offered. Consider a hypothetical variety store chain that chose to locate its first ten stores in affluent trade areas. Based on the demographic profile of its customer traffic (its "customer demographics"), this retailer might erroneously conclude that the concept has a high-end appeal. While this is not the case for variety stores, the "deck has been stacked" in favor of high-end customers as a result of the consistently affluent trade areas in which the retailer has elected to locate.

Another example could be a home improvement store chain with three stores located in the Bronx, a densely developed portion of New York City. Typically, the "target customer profile" for a home center chain consists of single-family homeowners—that is, consumers who own and reside in single-family homes and generally make more purchases in home improvement stores than consumers who rent high-rise apartments. A customer survey of our Bronx-based chain would probably reveal that most of its customers (or "customer demographics") are apartment renters rather than single-family homeowners. This result does not contradict the "target customer profile"—it simply reflects the fact that the propensity of single-family homeowners to spend more than renters in home improvement centers is, for the Bronx home improvement center chain, outweighed by the overwhelming number of renters (and relative lack of single-family homeowners), residing near the stores.

Relying solely on customer demographics to identify future store development opportunities could result in bad real estate decisions. Consider the variety store example cited earlier in this chapter; if only "customer demographics" were considered when screening for new store opportunities, the result would be the identification of locations in predominantly high-income areas, which is likely the opposite of the most appropriate strategy. While knowing a store's customer demographics can result in misguided real estate decisions, this information, when properly applied can be of considerable value. More specifically, knowing the demographic composition of each store's customers allows merchandisers to ensure that each store's merchandise mix is targeted to address the specific needs of consumers residing in each store's trade area.

Unit of Analysis

The identification of either a retailer's "customer demographics" or "target customer profile" begins with the identification of the most appropriate unit of analysis. Most retailers or service companies look at sales potential on a "per capita" basis; that is, the potential sales available for a particular area are a function of the number of people living there (as well as their demographic profile). Other retailers, such as home improvement centers and lawn and garden stores, sell merchandise that is primarily intended for a household rather than an individual; for those operators, households generally represent the most appropriate unit of analysis. A few retailers should most appropriately use a different starting point. The demand for auto supply retailers is primarily based on the presence of automobiles rather than people or households—as a result, registered automobiles would be an appropriate unit of analysis. Many office supply chains and computer superstores today sell as much or more merchandise to businesses than they do to individuals—for these operators, businesses represent the most appropriate unit of analysis for the business portions of their sales.

Customer Survey and Sampling Considerations

The most accurate method for determining the geographic area from which a retailer's customers originate would be to obtain the addresses and purchase amounts for every customer at a store over a one-year period. By using a one-year period, there would be no concerns relating to the fact that a store might draw customers from a greater distance during one time of year than during other times. Furthermore, given the number of individual customer transactions that would be included (which could be well over 1,000,000 for a high-volume operator such as a supermarket), the statistical level of confidence in the results would be virtually infallible.

Some retailers are fortunate enough to be able to do just that. Video stores which require membership for tape rentals inherently have the address and amount spent by every customer—as a result, a video store chain can generate an appropriate customer sample simply by tapping into its preexisting database. However, most retailers are not so blessed and find that the cost, or operational inefficiencies resulting from collecting every customer address over a one-year period are too high. For these retailers, the solution is to collect a representative sample of customer addresses and purchase amounts which contain enough observations to obtain an adequate level of statistical confidence, while concurrently keeping costs and operational inefficiencies to a minimum.

A less desirable source for assembling customer information is to utilize a preexisting database that includes information for a distinct subset of the retailer's customer base. For example, a department store might have detailed address information on its private label credit card transactions, or a drugstore chain might have similar information for its prescription customers. While it is very tempting to use these databases, it is critically important to first determine how large a proportion of total store sales they represent, and whether they are representative of the overall customer base. The department store chain might have marketed its credit card only to its most affluent customers—as a result, any customer analysis based strictly on credit card information will inherently overstate the importance of the department store's affluent consumers. Conversely, a drugstore's prescription customers might include a disproportionate number of elderly persons relative to those customers who shop the balance of the store. In such instances, it is important to determine whether the distribution of customers implied from the database sample is representative of the overall distribution of customers. This is usually accomplished by conducting a survey among a representative sample of customers in several stores, and then carefully comparing the customer distribution results against that implied by the database sample.

Contemporary POS systems can also be used to collect customer residence information. If the most appropriate unit of geography for a retailer is ZIP Codes (refer to Chapter 3), then customers can be asked for their ZIP Code at the point of sale. This ZIP Code collection program can be "turned on" for a predetermined period of time, and the resultant information used as the basis for a customer analysis. Some database companies are experimenting with collecting telephone numbers or other unique identifiers to pinpoint addresses down to smaller units of geography such as census tracts, which could be very helpful to retailers such as supermarkets and drugstores which typically serve smaller trade areas.

It is important for a retailer to determine the time of day and days of the week over which the sample is to be collected. In the case of large supermarkets or discount department stores, it might be possible to collect several thousand customer transactions during a single day. Unfortunately, such a limited time frame would inherently fail to incorporate any variations in customer distribution patterns that take place over the course of the week (i.e., between weekdays and weekends). Generally, the distribution of the sample should mirror actual sales distributions for an individual store—if 24 percent of a store's sales typically occur over the weekend, then 24 percent of the sample should also be collected over the weekend. The same concerns also apply to time of day—if a quick service restaurant conducts 50 percent of its weekday business during lunch hours (10:00 a.m. – 2:00 p.m.) and 25 percent during dinner (5:00 p.m. – 11:00 p.m.), then its sample should also be distributed accordingly. Table 4-1 provides an example of a recommended sample allocation for a quick service restaurant.

TABLE 4-1
Thompson Restaurants #305
Ann Arbor, Michigan
SAMPLE INTERVIEW ALLOCATIONS

	Drive-Thru				In-Store			
Time	Thurs.	Fri.	Sat	Sun.	Thurs.	Fri.	Sat.	Sun.
7 a.m. – 10 a.m.	21	21	10	4	11	11	4	3
10 a.m. – 2 p.m.	72	72	28	17	78	78	32	18
2 p.m. – 5 p.m.	26	26	10	7	18	18	5	5
5 p.m. – 8 p.m.	33	33	13	10	19	19	7	8
8 p.m. – 11 p.m.	18	18	6	3	7	7	2	2
Daily Totals	170	170	67	41	133	133	50	36
Total Interviews				448				352

	Drive-Thru		In-Store	
	Sales	Number of Interviews	Sales	Number of Interviews
Total Sales	56%	448	44%	352
Sales by Weekday	76%	340	76%	266
Sales by Weekend	24%	108	24%	86

The time of year can also be a consideration in selecting an appropriate sample. The Christmas season often can result in shopping patterns that are atypical relative to the balance of the year—consequently, conducting a one-week sample starting on December 17 may result in some misleading findings. Conversely, for a retailer whose sales are heavily dependent on the Christmas season, this may represent an appropriate time to collect a customer sample. The most appropriate time of year to collect a customer sample can also depend upon the merchandise being offered. Nurseries are an example of a retailer whose sales show significant increases during the spring months—conducting a customer intercept survey for a Manitoba nursery chain in January could be very misleading.

The logistical method of collecting a customer sample for the purpose of determining a customer profile is an important consideration for retailers, since it can significantly influence both the cost and accuracy of the sample collection procedure. There are four primary questions that have to be considered:

1. What unit of geography (i.e., ZIP Code) should I be collecting?
2. Is the customer sample representative of my overall customer base?
3. What types of customers do I need to consider?
4. How many customers do I need to collect information from?

The first point is covered in detail in Chapter 3. The second consideration can be relatively straightforward: Can the customer survey reach all of the potential customers who are shopping a store? Drive-thru portions of quick-service restaurants can account for over one-half of total sales—a survey conducted among only those patrons who park their vehicles and go inside the restaurant to eat or pick up their food would therefore be inherently incapable of reflecting the overall customer base. Some lumberyards have special checkouts that are designated only for contractors—a failure to allocate some of the sample to those checkouts will inherently under represent the importance of contractors to the overall customer base.

The third consideration constitutes the types of customers who are patronizing the store in question. For office supply stores, it is important to ascertain whether the customer is shopping for personal use or for his or her place of business; for quick-service restaurants, customers should be asked whether they are coming from home, work, a shopping trip, or just happen to be driving by on a major highway. The address or unit of geography collected may also vary based upon the reason for the trip. In the fast food example, the customer's home address is important for customers who are coming from home, but not for those coming from work or a shopping area. Similarly, a contractor patronizing a lumberyard may be doing so due to its proximity to his or her job site rather than proximity to home.

The fourth and final consideration relates to the number of customers that need to be included in the sample. There are a number of sources that can be consulted to determine how large a sample should be to achieve a given confidence level. The comments that follow are not intended to be a precise "blueprint" for calculating the required sample size for any situation, but merely provide general guidelines and cautions for retailers.

The fundamental rule regarding sample size is the more the better. The statistical reliability of a sample is always dependent on the number of observations in the sample. It is important that a customer sample be large enough to ensure that an ample number of observations are collected for each trade area sector (i.e., each ZIP Code or census tract). For example, it would not be advisable to generalize about how a store's customers and sales are distributed throughout its trade area by trade area sectors (i.e., ZIP Codes), if the sample originating from each sector is comprised of only a few customers or transactions. Generally, for high-traffic stores with a trade area comprised of 15 to 30 units of geography (ZIP Codes or census tracts), a sample of at least 1,000 customers is desirable. With lower-traffic stores that have relatively few transactions, it may be necessary to work with a smaller sample (perhaps 400 to 600 transactions).

Customer Demographics

It is relatively easy to determine a retailer's customer demographics by simply "piggy backing" a few pertinent questions concerning demographics along with customer origin surveys (e.g., customer exit surveys, POS surveys). These surveys are usually administered by checkout personnel or an independent interviewer. The nature of the demographic information collected is usually a reflection of each retailer's respective business. For example, home centers would likely be interested in collecting housing information, whereas toy stores may be interested in collection information pertaining to family composition. Once a sample of customers has been surveyed, the retailer can identify its customer demographics by simply aggregating the information collected in the customer survey, and then calculating median and average values where appropriate.

Target Customer Profile

A variety of statistical techniques can be used to identify a target customer profile once the customer sample has been collected. To accomplish this, one must first calculate the "sales penetration level" that a store is achieving in each surrounding sector (refer to Chapter 3 for a description of how to calculate sales penetration levels). To better understand how a target customer profile is calculated, an example using one store is presented. This is an oversimplified example, which is presented here merely to illustrate the issues involved in identifying a target customer profile. In practice, a retailer should include as many stores as is practically possible within the database.

Figure 4-1 displays the sales penetration level (average annual sales per capita) being achieved in each ZIP Code surrounding a store, as well as the driving distance from the center of each ZIP Code to the store, and the median household income level within each ZIP Code. As a general rule, sales penetration levels are lower in ZIP Codes located further away from the store. This phenomenon, commonly referred to as "distance decay," reflects the fact that inconvenience and competition typically make it less likely that consumers residing farther away from a store will shop there.

figure 4—1

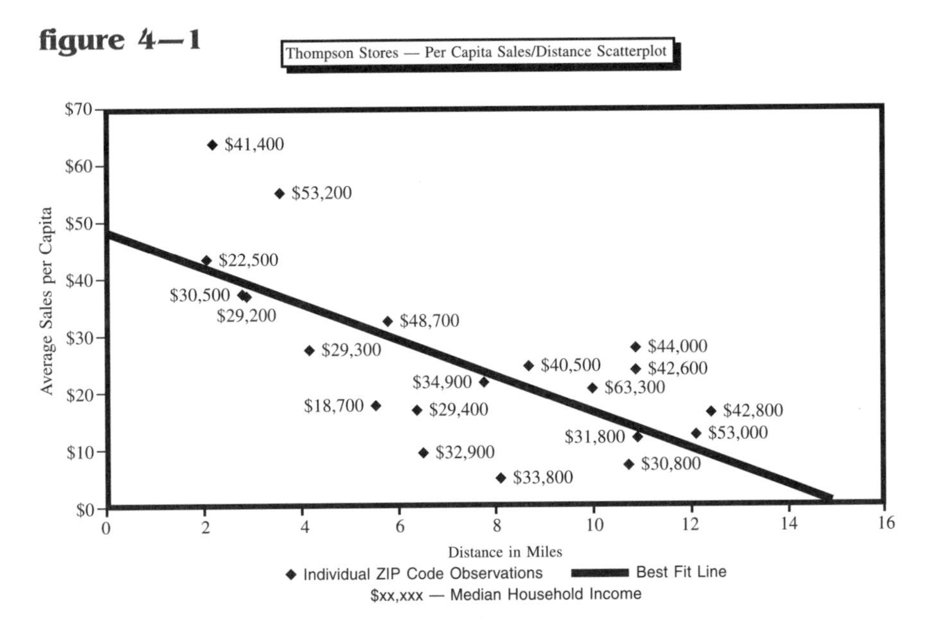

Regression analysis is one of the most widely used techniques for measuring the strength of the relationship between two variables. The "best fit" line shown in Figure 4-1 provides a linear expression of the relationship between distance traveled by consumers and their respective sales penetration levels. The mathematical basis for defining the "best fit" line is that the square of the distance from the line to all of the points plotted (which is the information upon which the line is based) must result in a mathematical minimum when summed. By this criterion, the resulting line represents the "best fit" for the underlying data, as it is closest to all of the points.

While the "best fit" line illustrates the overall relationship between distance and penetration levels, it is clear that some of the ZIP Code observations are well above the line (indicating that they are performing much better than expected given their distance from the store), and others are located below the line (they are performing much worse than expected given their distance from the store). The identification of a target customer profile, in a nutshell, consists of identifying those demographic characteristics that are consistently present in trade area sectors that are overperforming relative to their distance from the store. In the example presented in Figure 4-1, all ZIP Code observations located above the line (i.e., which are experiencing better than average sales penetration levels given their distance from the store), have median household income levels of $40,000 or higher, while all ZIP Codes which are underperforming, or located below the line, have median household income levels below $40,000. As a result, the target customer for this retailer would have household income levels of $40,000 or greater.

The actual relationship that exists between sales penetration levels and demographic characteristics is invariably much more complex than the preceding example. To account for this, it is generally appropriate to simultaneously test the correlation between sales penetration levels and a wide variety of demographic factors, such as age, occupation, education, and so forth. Multiple regression analysis represents one statistical technique that can be used to accomplish this. The result of this procedure is typically a series of demographic characteristics that are positively correlated with sales performance levels. In the preceding example, multiple regression analysis might reveal that the following demographic variables comprise the "target customer profile" for the retailer in question:

- Median household income over $40,000
- High proportion of college graduates
- High proportion of white-collar employees

The retailer would therefore be well-advised to search for future locations that have concentrations of affluent, highly educated, white-collar consumers in their immediate vicinity. While this information is certainly helpful, there are additional insights that can be gleaned from it. Of the three variables in question, which is the most important? If the retailer has a choice between two potential locations—one surrounded by highly educated consumers with modest incomes, and the other surrounded by affluent blue-collar consumers with high school educations, which will result in the highest potential sales levels?

Multiple regression analysis is a technique that takes the interrelationship between variables into account, and allows the retailer to focus on the one or two variables that are the most important. Multiple regression analysis identifies the combination of variables that collectively are most positively correlated with sales penetration levels. In this way, stepwise regression analysis allows the analyst to evaluate potential concerns of autocorrelation, or the interdependency between two demographic variables. In the preceding example, it may be that education is really the driving variable behind the success of the retailer. However, because education and income are autocorrelated (i.e., well-educated people tend to make more money), the initial regression analysis identified a positive relationship between income and sales performance levels, as well as between education and sales performance levels. By using multiple regression analysis, a retailer would be able to determine that education is the variable that is most positively correlated with sales, and would select the site surrounded by highly educated consumers with modest incomes rather than the site surrounded by affluent blue-collar consumers with high school educations.

5 Competitive and Site Characteristics Analysis

THE PREVIOUS CHAPTER FOCUSED ON THE RELATIONSHIPS that exist between consumer demographic characteristics and retail sales performance. Two other factors that affect sales performance levels are competition and site characteristics.

Analyzing the impact of competition on store sales performance is more involved than merely determining whether a particular retailer is "competitive" or not. Important considerations include determining which retailers are directly versus indirectly competitive; whether there are regional variations in a competitor's strength; whether being the "first one in" a market, impacts the relative strength of a competitor; and how a competitor's location relative to a store affects its competitive impact. Is a retailer better off locating across the street from a competitor, or as far away as possible? A careful analysis of competition will not only reveal which competitors have the greatest effect on sales performance, but will also provide guidance as to how best to locate within a market so as to neutralize or mitigate the impact of competition.

Site characteristics can also strongly influence the sales performance ultimately achieved by a retailer at a specific location. By understanding the relationship that exists between site characteristics and sales performance, a retailer can determine, for example, whether there is an acceptable cost benefit associated with the higher rents typically charged for sites with more favorable site characteristics. Some retail concepts are critically dependent upon clear visibility and ready access to the site in order to achieve successful store performance, while for others, these factors may have relatively little impact on sales performance. Furthermore, the importance of site characteristics may also be dependent upon the presence of strong nearby competition; that is, site characteristics may be of little consequence to a store's performance unless a competitor takes a nearby location with clearly superior characteristics (e.g., superior visibility, parking, etc.).

The first half of this chapter addresses the approaches taken to quantify the impact that competition has on store performance. The second half of the chapter is devoted to assessing the degree to which site characteristics impact a retailer's sales performance.

COMPETITION

Competitive Inventory The first step in evaluating the impact of competition on the performance of database stores is to identify all potentially significant competitors surrounding each of the database stores. In most large markets there will be many stores selling competing lines of merchandise, but fewer representing directly competitive formats. In some instances, small units that are not directly competitive represent "noise" and will not have a quantifiable impact on store sales performance; in other instances, their impact may be significant. Beginning with an appropriate inventory of potential competitors is critical, for the decision to omit a certain retailer makes it impossible to measure its impact on store performance—if truly competitive, this retailer might be inadvertently overlooked when evaluating future locations. On the other hand, collecting information on hundreds of tangential competitors that are likely to have no significant impact on database store performance can be time-consuming and costly.

Ultimately, the analyst should rely on common sense and personal experience to determine which competitors to consider. For example, if a competitive analysis were being conducted for a conventional supermarket chain, it would clearly be appropriate to include all other conventional supermarkets. The supermarket portion of a discount department store/food store retailer such as Wal-Mart Super Center would also be included. Warehouse membership clubs, such as Price/Costco and Sam's Club, are generally not directly competitive with a supermarket, but might warrant inclusion as "indirect" competitors. On the other hand, it would probably be inappropriate to include convenience stores in a competitive analysis. While convenience stores sell many of the same food items as supermarkets, their individual sales and competitive impact on a store by store basis is so small as to be immeasurable. Furthermore, it is highly unlikely that a supermarket chain would choose to alter its location strategy simply because a convenience store happened to be located nearby.

Once the inventory has been conducted for each database store, the next step is to categorize competitors into three broad categories, as follows:

> **Sister Stores** — Sister stores (other stores in the same chain) represent the most direct and comparable competition to the database stores. As an example, a consumer would generally be unlikely to drive past one Sports Authority sporting goods store to get to another, given that the two stores offer the same merchandise at the same price.
>
> **Direct Competitors** — Direct competitors are retailers whose underlying format and concept are similar to the retail chain in question. For example, both Jumbo Sports and Sportmart represent direct competition to The Sports Authority—all three chains are large (30,000 to 50,000 gross square feet), typically seek nonmall locations, and carry a broad assortment of sporting goods and athletic footwear and apparel. There may be differences between operators (e.g., one may have a lower price image, or offer a broader selection of team sports equip-

ment), but the underlying concept is similar. While a consumer would generally not drive past one Sports Authority unit to get to another, he or she might bypass a Jumbo Sports or Sportmart to shop at Sports Authority.

Indirect Competitors — Indirect competitors carry some or all of the same merchandise lines as a retailer, but the overall selection or underlying retail concept is fundamentally different. As an example, warehouse membership clubs sell food, but represent indirect competition to conventional supermarkets because they require a membership to shop there, only carry bulk quantities of food at significantly discounted prices, and carry only a small fraction (perhaps 5 percent to 10 percent) of the stock keeping units (SKUs) available at a conventional supermarket.

Relative Competitive Locations Having conducted an inventory of competitors who potentially affect each database store, their locations should next be plotted on each database store's trade area map. Then, for each sector (assumed to be ZIP Codes for purposes of this discussion) within the trade area of each database store, one must determine the geographic relationship that exists between the database store and each competitor. For every trade area sector, each competitor is classified as being either:

- **Adjacent** — Competitor is located adjacent (or effectively adjacent) to the database store.
- **Intercepting** — Competitor is in a position to intercept, or cut off, all or most of the consumers in a particular ZIP Code that would otherwise shop the database store.
- **Impacting** — Competitor exerts an impact on a particular ZIP Code, but is not in a superior position to serve that ZIP Code relative to the database store in question.

Map 5-1 provides an example of each of the three competitive impact classifications described above. The ability of Goodrich Store #95 to attract sales from ZIP Code 66216 is affected by three competitive facilities: Stay-Low Stores is adjacent, Greg's is intercepting, and Caroline Stores is impacting. The competitive environment is less intense for ZIP Code 64113; Stay-Low Stores is still adjacent (as it is for every ZIP Code in the trade area) and Caroline Stores is impacting, but there is no intercepting competitor. This process of categorization is carried out for every ZIP Code in each database store's trade area, and the resultant calculations are incorporated into a partial correlation or regression analysis.

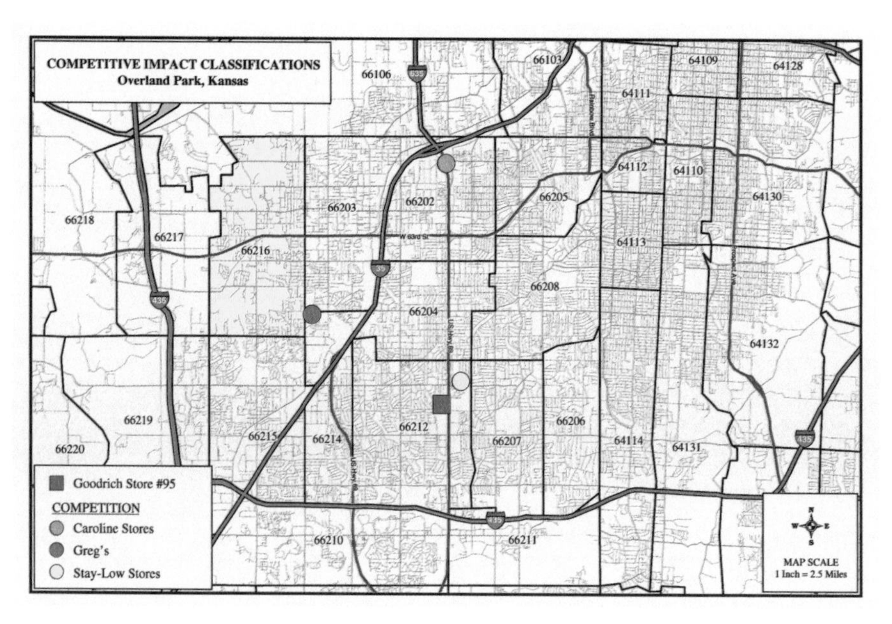

map 5—1

This analysis measures each competitor's relative strength. A formidable competitor will impact a database store's sales performance from individual ZIP Codes no matter whether it is in an adjacent, intercepting, or impacting position. In other words, it will not need the advantage of relative locational convenience in order to exert a strong competitive impact. A weaker competitor, however, may only have a discernible impact if it is located in an intercepting position, where it can use a significant locational convenience advantage to overcome any perceived operational or merchandising deficiencies. A significant disadvantage of this method of competitor measurement can be the difficulty of distinguishing between intercepting and impacting. Fortunately, the increased use of GIS technologies in conjunction with retail sales forecasting has made such measurements more consistent. However, the measurement algorithms to assess intercepting and impacting competition can become quite complex.

The intercepting/impacting/adjacent scheme has proven to be an effective approach, but is by no means the only possible or useful competitive measurement system. A retailer should test a number of different coding schemes using a variety of measurements such as count, sales area, or sales if available. The optimal approach to measuring the impact of competition on retail performance is likely to be as varied as the retailers for which it is used.

The statistical techniques used to measure the relative impact of competition are the same as those described in Chapter 4. Map 5-2 and Figure 5-1 provide a graphic example of the relationship between competition and sales performance. In this example, there are two competitors affecting Amon Stores #103—Kinnick Stores and Long's. Kinnick Stores is located in an intercepting position relative to ZIP Code 30342, and Long's is in an intercepting position relative to ZIP Code 30084. In both cases, the performance of the ZIP Codes in question is below the "best fit" line, suggesting that those competitors are having a significant impact on sales performance levels. There are also two ZIP Codes that are impacted (but not intercepted) by competition—ZIP Code 30346 is impacted (but not intercepted) by Kinnick Stores, while ZIP Code 30340 is impacted (but not intercepted) by Long's. However, an examination of the "best fit" line reveals a dissimilar impact. ZIP Code 30346 is situated below the "best fit" line, suggesting that Kinnick Stores affects the sales performance of Amon Stores #103 even when it does not have the benefit of relative location convenience. ZIP Code 30340, however, is actually situated slightly above the "best fit" line, suggesting that Long's does not impact Amon Stores #103 unless it is more convenient to the consumers. The implication of this

map 5—2

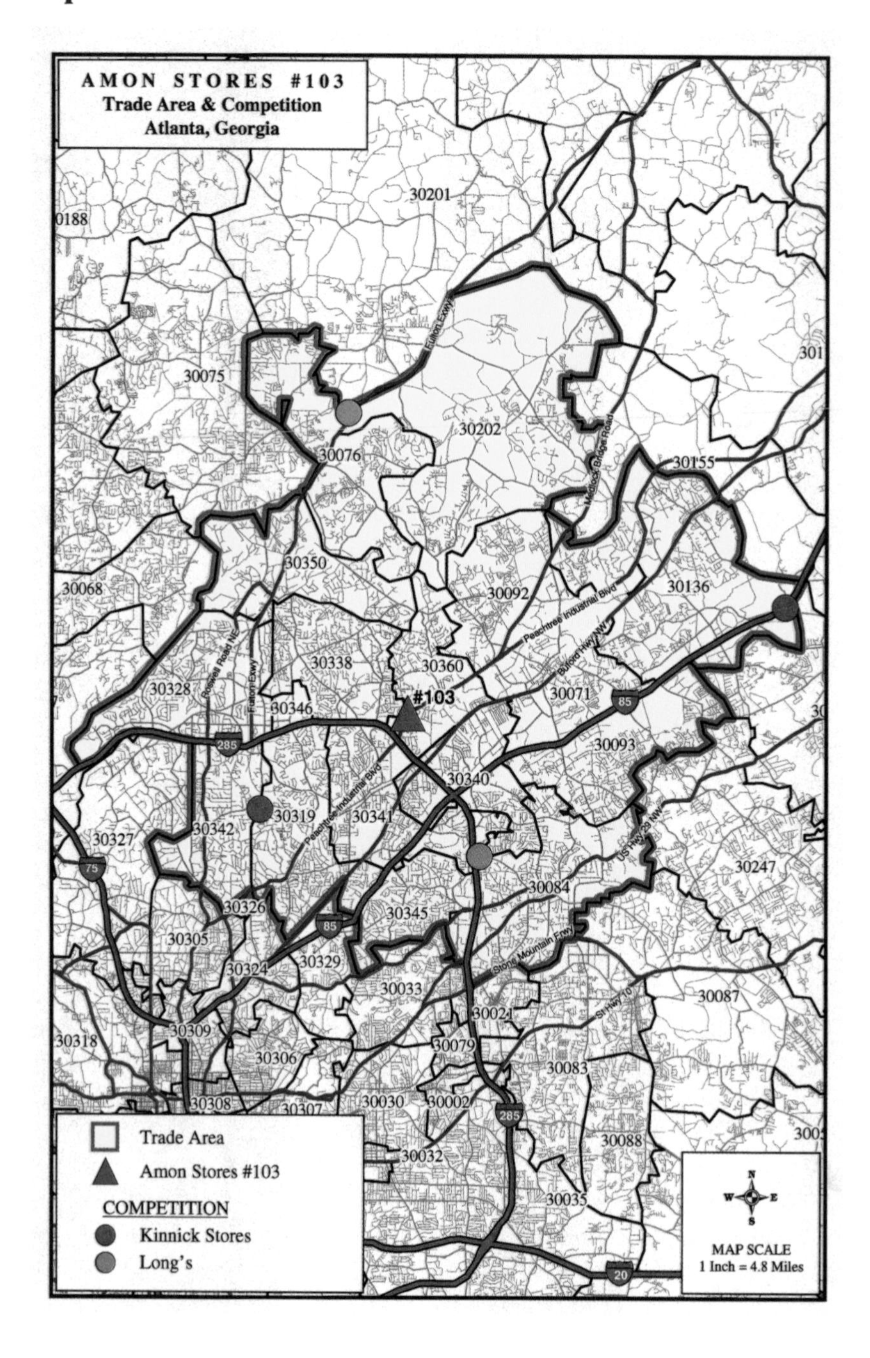
AMON STORES #103
Trade Area & Competition
Atlanta, Georgia
30201
30075
30076
30202
30155
30350
30092
30136
30068
30338
30360
#103
30328
30346
30071
30093
30340
30319
30341
30342
30327
30084
30247
30326
30345
30305
30324
30329
30033
30087
30309
30318
30306
30079
30083
30308
30307
30030
30002
30088
30032
30035
Fulton Exwy
Peachtree Industrial Blvd
Buford Hwy NW
Roswell Road NE
US Hwy 29 NW
Stone Mountain Frwy
Trade Area
Amon Stores #103
COMPETITION
Kinnick Stores
Long's
MAP SCALE
1 Inch = 4.8 Miles

figure 5—1

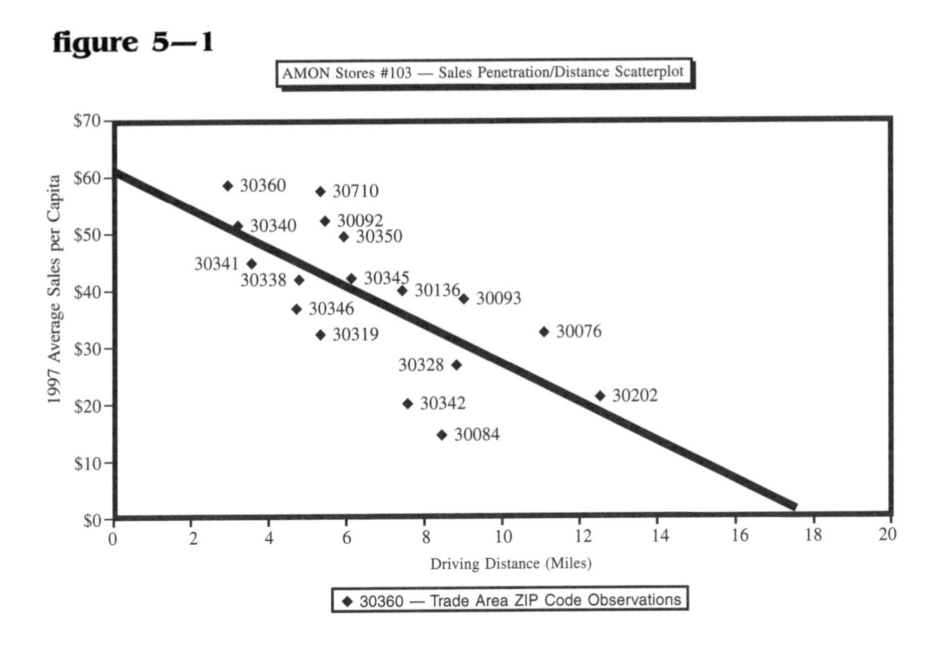

analysis would be that Kinnick Stores is a stronger competitor than Long's.

The preceding discussion represents an oversimplified example for measuring competitive impacts. In reality, it is helpful to have dozens of observations for each competitive scenario (e.g., Kinnick Stores in an intercepting position, Kinnick Stores in an impacting position) to develop as much confidence as possible in the outcome. The statistical techniques (such as regression analysis) used to measure competitive impacts can provide insights by simply distinguishing between intercepting and impacting. By evaluating the statistical strength of the relationship between competitors in different locational situations and their corresponding impact on sales performance, a retailer can quantify and directly compare the competitive impacts of different chains.

An evaluation of the impact of adjacent competitors can sometimes produce what initially seem to be counterintuitive results. Some retailers actually benefit from having competition next door—that is, the presence of an adjacent competitor results in higher sales than if the adjacent competitor did not exist. This is most frequently the case for comparison goods retailers such as mall jewelry stores—the presence of several jewelry operators under one roof produces a greater retail attraction and can result in increased sales for everyone. In other words, the whole really is greater than the sum of the parts. For others (particularly convenience goods retailers), adjacent retailers generally result in lower sales. Here, the impact of an adjacent retailer is a direct test of the relative strengths of the two operators (with respect to their merchandising, pricing, service, and so on), as neither operator offers consumers the advantage of relative locational convenience in order to "lure" customers away from the other.

The results of a regression analysis to determine the impact of competition on sales performance can be of value in helping a retailer formulate a deployment strategy designed, to mitigate the effectiveness of competitors within a market. For example, if the regression analysis reveals that a retailer is a significantly weaker operator than one particular competitor, then a strategy of locating between (rather than next to) that competitor's stores might be an appropriate one, in order to provide consumers with a compelling reason to shop the retailer rather than the competitor (e.g., the store has the advantage of locational convenience for at least a portion of the market). Conversely, if a retailer is much stronger than its competitor, then all other things being equal, it would be advantageous to locate next to that competitor, thereby denying the competitor any locational convenience-related advantages.

Unique Competitive Considerations There are a number of other issues that should be evaluated when assessing the strength of competition. One such issue is that competitors rarely "stand still" with respect to their merchandise mix, operating policies, physical plant, and so on. Therefore, as with other market characteristics, the impact of competition on store sales performance should be monitored over time. Changes in merchandising, operations, and marketing strategies can alter the impact that specific competitors have on a retail chain. For example, when large-format office supply chains were in their infancy in the mid to late 1980s, their competition initially consisted primarily of independent "mom and pop" office supply operators, and indirect competitors such as discount department stores. However, with the ongoing maturity of the large-format office supply store as a distinct retail classification, competition today consists primarily of other large-format office supply stores.

Competitors can also exhibit significant regional variations in their performance; it is common for a retailer to generate higher sales in some parts of the country than in others. There can be a variety of explanations for this, including merchandising sensitivities to different regions of the country, the relative size and condition of facilities, management strengths and weaknesses, the strength of competition in different regions, and the like. Whatever the underlying reason, one result of these competitive "inconsistencies" is that a retailer may need to adopt alternate deployment strategies against a competitor by region—perhaps going head-to-head where possible in one region, while simultaneously avoiding sites that are adjacent to the competitor in others.

The age or maturity of competition relative to the retailer in question can also affect the degree of their impact. For some retailers, being the first operator in a new market can significantly enhance their position. As an example, the first warehouse membership club to locate in a market will often develop a loyal consumer and business customer base, making it difficult for subsequent operators to lure customers away. Conversely, dinnerhouse restaurants often perform in exactly the opposite manner—consumers are often eager to try a new operator, with the result that a new restaurant often does better in the first few weeks and months after opening than the second or third year of operation. A statistical analysis can help identify the relative importance of length of time in the market, thereby allowing a retailer to better understand the importance of factors such as getting into the market first, maturation, and so on.

Site Characteristics

Site characteristics, or the physical attributes of the site occupied by a retailer and the immediate surrounding area, can have a significant impact on the sales performance of a retailer. As a general rule, the success of convenience-oriented retailers, such as supermarkets, drugstores, video stores, and dry cleaners, is highly dependent upon the quality of the site. For example, a video store located at a site with obstructed visibility or difficult ingress/egress relative to a nearby competitor would likely have difficulty in capturing a significant share of the sales potential within its trade area. Poor site characteristics are usually less of an impediment to the sales performance of comparison goods retailers, as their appeal is more dependent upon merchandise selection and quality, price points, and customer services, rather than convenience issues such as parking, and ingress/egress. Because of this, many comparison goods retailers can occupy facilities with less than ideal site characteristics, often with little or no diminution in sales performance. As an example, a shopper may bypass several conveniently located jewelry stores in favor of a jeweler with an excellent reputation for quality and service. Such jewelers can be located in obscure sites with no measurable impact on their sales performance because of the strong comparison goods nature of their business.

There are several distinct factors that fall under the general category of site characteristics, as follows:

> **Visibility** — The ease with which passing motorists or pedestrians can see the retailer. Visibility affects the ability of a first-time shopper to find the store, as well as the "top-of-mind awareness" that can result from having a highly visible location.
>
> **Parking** — The number of available parking spaces proximate to the entrance of the retailer. In more urbanized areas, the availability of free surface parking, as well as the proximity of mass transit stops, can also be a consideration.
>
> **Ingress/Egress** — The ease with which a motorist can get into and out of the store parking lot. Considerations include the presence of a left-turn

lane, signalized access, the number of ingress/egress points, median breaks, deceleration lanes, and traffic congestion in the immediate vicinity of the site.

Accessibility — The regional accessibility associated with a site, or the ease with which consumers throughout the trade area or market can get to the site.

Retail Synergy — The presence (or absence) of other retailers in the immediate vicinity of the site which may help attract consumer traffic to the area.

Safety/Security — The perceived safety and security associated with the site and the immediately surrounding area. It is important to note that local perceptions of safety are much more important than actual crime statistics, since it is perceptions that will influence actual shopping behavior.

The same statistical techniques described earlier in this book to evaluate the impact of consumer demographic characteristics and competition on retail sales performance can also be used to evaluate the importance of site characteristics. However, the number of independent observations is limited to the number of stores in the database. That is, each database store provides only one observation per site characteristic being tested (e.g., parking, visibility, etc.), so that a 40-store database provides a sample of only 40 observations. Conversely, if each database store has an average of 15 ZIP Codes within its trade area, the same 40-store database would yield a total of 600 observations for purposes of evaluating the relationship between demographic characteristics and sales performance, resulting in a much higher level of statistical confidence.

It is also important that the procedure used to quantify the site characteristics associated with each database store be as objective and consistent from store to store as possible. As an example, it would be inappropriate to allow the manager of each database store to evaluate the characteristics of his or her own store, since each manager's benchmark, or yardstick for evaluating site characteristics is different. Ideally, the analyst developing the database should visit each database store to evaluate their site characteristics—if this is impossible or cost-prohibitive, then a carefully worded description of site characteristics ratings should be drafted and disseminated to appropriate parties to make the results as directly comparable as possible. Figure 5-2 is a sample site diagram form; associated instructions are presented in Figure 5-3.

FIGURE 5-2
RICHARDS DRUGS
SITE EVALUATION FORM

Background Information	
Store #	
Street	
Cross-street	
City	
State	
Analyst Name	
Date	

Site Characteristics	**Excellent**	**Good**	**Fair**	**Poor**
Visibility				
Parking				
Ingress/Egress				

Site Diagram

A thorough site analysis will enable a retailer to quantify the importance of site characteristics on sales performance levels, thereby providing guidance as to which site characteristics can be compromised in future location opportunities, as well as which are of critical importance. Furthermore, the analysis may also yield insights into the importance of site characteristics in the face of strong competition—it may be that the importance of favorable site characteristics becomes even more significant when there is a strong

FIGURE 5-3
SITE EVALUATION FORM INSTRUCTIONS

Please fill out 3 sections of the Site Evaluation Form (Figure 5-3) as follows:

1. **Background information**: Please fill in all necessary information.
2. **Site characteristics**: Please rate the site "Excellent," "Good," "Fair," or "Poor" for each of the four characteristics listed below. The criteria needed to determine which rating to use are as follows:
   - Visibility
     - **Excellent** — This rating indicates a nearly flawless site.
     - **Good** — This rating indicates the facility is either clearly visible from most streets or gets direct daily exposure from a larger customer segment, with only a few minor flaws.
     - **Fair** — This rating indicates the site may be visible from certain streets, and also possesses some significant limitations for daily exposure to potential customers.
     - **Poor** — This rating indicates a site that is difficult to see from most streets and gets limited daily exposure from potential customers.
   - Parking
     - **Excellent** — This rating indicates that convenient, nearby parking is always available to customers.
     - **Good** — This rating indicates that parking is usually, but not always, available near the store.
     - **Fair** — This rating indicates that it is often difficult to Find available parking near the store.
     - **Poor** — This rating indicates that it is usually very difficult for customers to find a parking space anywhere near the store.
   - Ingress/Egress
     - **Excellent** — This rating indicates that the site is easily accessible from several ingress/egress points along all of the thoroughfares that it abuts, with signalized access or designated left-turn lanes, if necessary.
     - **Good** — This rating indicates that it is usually easy to access the site, but may be difficult from certain access points or at certain times of day.
     - **Fair** — This rating indicates that it is often difficult for a motorist to access the site from the surrounding road network, due to heavy traffic or inadequate access points.
     - **Poor** — This rating indicates that it is usually very difficult for a motorist to access the site.
3. **Site Diagram**:
   Either draw a site diagram depicting the site in relation to nearby roads, shopping center(s) and other land-use types, or attach a recent aerial photograph with the site clearly delineated.

competitor with good site characteristics located nearby. In this way, a retailer can be more discriminating when evaluating the relative viability of alternative locations, particularly as it relates to the willingness to pay a higher rent for a superior site.

6 Prioritizing Markets

THE DEVELOPMENT OF A STORE DATABASE AND THE associated customer and competitor analysis described in the preceding chapters, provides a retailer with the necessary information from which to begin strategic analysis of new markets and sites for expansion. One of the most useful analytical processes that can be implemented after a store database is developed is to prioritize prospective markets according to their viability for the retailer. The process of market prioritization (or market screening) involves identifying markets that have the greatest concentration of in-profile customers and weakest competition, as these usually offer the most attractive expansion opportunities for a retailer. Many retailers look at market prioritization studies much like an insurance policy; such analysis enhances the likelihood that markets chosen for expansion will have the requisite ingredients to achieve successful sales levels for the chain. Additionally, prioritizing markets based on their demographic and competitive attractiveness enables a retailer to determine if the most viable markets are concentrated in a particular region of the country, thereby facilitating a logical store deployment plan. Ultimately the goal of a comprehensive market screening is to allow a retailer to prioritize markets for new store development so that the retailer may focus on those which are most likely to yield favorable store performance.

This chapter will focus on screening markets according to their unique demographic and competitive composition. This is not to suggest that economic factors that have traditionally influenced the market screening process (e.g., land costs, unemployment rates, the labor pool, and advertising costs) should be ignored. Rather, the market screening process should incorporate demographics and competition as well as traditional economic factors. A screening process which combines both approaches is most likely to reveal the best markets for expansion opportunities.

Before addressing the mechanics of the market screening process, a word of caution. It is tempting for retailers to presume that the largest markets in the country represent the most attractive expansion opportunities. However, as demonstrated in the preceding chapters, most retailers have a distinct and measurable customer profile; therefore, prioritizing markets based on density alone does not account for the attractiveness of their demographic composition or the intensity of their competitive environment. More specifically, a smaller market with a significant concentration of in-profile consumers may yield greater sales potential than a denser but less in-profile market.

Every retailer poised for expansion must first identify the geographic area in which expansion is to occur. As an example, a fledgling retailer will likely not have the resources to embark on a national expansion campaign. Such a retailer would likely want to limit the scope of a market screening study to a region of the country, or even to a specific state or states. Conversely, more mature retailers would likely want to identify the best expansion opportunities nationally. For the purposes of this chapter, a hypothetical bookstore retailer, "Thompson Stores," has determined that it has the distribution network, financial backing, and operational capability to pursue a national expansion strategy throughout the United States. Thompson Stores' next challenge is to determine which markets, beyond those in which it currently has a presence, offer the greatest potential for expansion.

Having developed a store database, Thompson Stores management knows that its customer profile includes people who are well educated (persons with at least a bachelors degree), are employed in white-collar occupations (professional, executive, and sales occupations), and have total household incomes of greater than $100,000 annually. The database development process also indicated that Thompson Stores realizes its strongest competitive impact from large-format book stores such as Borders Books & Music and Barnes & Noble.

The steps in undertaking a national market prioritization are as follows:

- Determine the minimum total population or households of markets that will be considered in the analysis. Markets with population or household totals that are inadequate to support a store successfully should be excluded from this analysis.

- Develop a logical grouping/segmentation of markets by size. The most accurate and equitable analysis of markets is to group them according to size classifications so that markets of similar population or households are being compared against each other.

- Collect and process competitive and demographic data for every market to be considered.

- Develop a viability score for each market based on key variables such as demographics and competition.

Determine the Minimum Density Threshold of Markets to be Analyzed

The first step in conducting a Market Prioritization analysis is to establish a minimum population (or household) threshold for the markets that will be considered in the prioritization process. More specifically, it is necessary to determine the minimum population density that is necessary to achieve acceptable sales levels for a store. This level of density then defines the minimum size of markets that should be considered in a market screening analysis.

Generally, a minimum size threshold can be determined by examining the population within the trade areas of database stores. This process usually begins by searching the database for the store that has the least populated trade area while achieving acceptable levels of sales performance. The trade area of this store represents a reasonable minimum size threshold when screening markets, as implicitly, no stores have demonstrated an ability to perform successfully with population below this level. However, the database store with the smallest trade area population base often has very favorable demographic characteristics and a weak competitive environment, thereby allowing the store to achieve acceptable sales levels with a smaller population base than the average database store's trade area.

Further examination of the store database will usually reveal that trade areas with strong competition or less favorable demographics need to have larger population bases in order to support acceptable sales levels. As such, the average trade area density exhibited by the entire set of database stores inherently reflects the best and the worst demographic and competitive situations encountered by the chain and, therefore, represents a good barometer as to the *typical* population threshold needed to support at least one store. Thus, the lowest population trade area in the database establishes an approximate minimum threshold needed to support a store under ideal circumstances, while the average trade area for all database stores provides an indication of the density needed to support a store in more typical situations.

An analysis of the hypothetical Thompson Stores database reveals the following. Table 6-1 indicates the database store with the lowest trade area population is located in Bradford, which has a trade area population base of 228,054 people, and exhibits the most favorable demographic composition of all 20 database stores. Further review of the Bradford store's performance reveals that 10 percent of the store's sales originate from consumers who reside beyond its trade area. To be conservative, the number of consumers estimated to be supporting the Bradford store has correspondingly been increased by 10 percent (to 250,000 people) to help account for sales originating from beyond the trade area. It can therefore be assumed that, under ideal conditions, a minimum of 250,000 people in a market is required to support one store. However, given the fact that the average trade area density for the database is 450,000 people, it is reasonable to assume that somewhere between 250,000 and 450,000 people in a market will be necessary to support a single store. Thus, the national market screening will not consider any market that has less than 250,000 people; further, markets between 250,000 and 450,000 people likely represent markets that could support one store at most.

TABLE 6-1
THOMPSON STORES
TRADE AREA ANALOG SUMMARY

Location	1997 Pop.	1997 Pop. Sq. Miles	1990 % Exec/ Prof/Sales	1990 % College Graduate	1997 % Income > $100K	1997 Median HH Inc.	Per Capita Sales	1997 Trade Area Sales	1997 Beyond Sales	1997 Total Sales	Capture Rate
Lompoc	765,630	3,060	60%	51%	23%	$63,896	$28.80	$22,052,968	$5,277,293	$27,330,261	80.7%
Sidney	725,341	4,971	42%	23%	10%	$41,841	$16.92	$12,273,107	$3,523,919	$15,797,026	77.7%
Hampton	602,541	2,915	46%	29%	19%	$56,445	$17.38	$10,471,986	$7,817,621	$18,289,607	57.3%
Salem	587,430	1,872	46%	31%	18%	$57,847	$17.16	$10,079,062	$3,101,497	$13,180,559	76.5%
Morburg	536,796	1,550	46%	32%	15%	$50,538	$28.24	$15,158,990	$6,447,434	$21,606,424	70.2%
Tipton	530,178	2,345	58%	46%	19%	$49,741	$23.67	$12,551,434	$4,724,307	$17,275,741	72.7%
Wayne	470,125	881	50%	30%	16%	$54,639	$32.67	$15,359,519	$6,270,131	$21,629,650	71.0%
Springfield	454,946	2,601	57%	46%	20%	$62,087	$26.23	$11,934,377	$2,645,482	$14,579,859	81.9%
Morley	445,131	1,833	59%	50%	27%	$72,358	$23.77	$10,579,009	$3,991,841	$14,570,850	72.6%
Oakland	437,024	2,810	45%	25%	13%	$49,738	$3.38	$1,475,435	$423,983	$1,899,418	77.7%
Lakeside	436,133	1,199	57%	42%	22%	$60,437	$4.07	$1,774,958	$679,902	$2,454,860	72.3%
Westland	425,552	1,733	54%	40%	14%	$55,460	$4.27	$1,816,869	$653,821	$2,470,690	73.5%
Lancaster	410,006	629	33%	18%	6%	$37,943	$4.41	$1,809,830	$467,307	$2,277,137	79.5%
Gaylord	403,790	847	46%	32%	15%	$58,601	$3.46	$1,397,394	$372,487	$1,769,881	79.0%
Wilmott	345,238	965	50%	37%	20%	$57,483	$5.80	$2,000,749	$678,812	$2,679,561	74.7%
Chesny	344,159	5,193	49%	32%	15%	$48,592	$7.22	$2,485,561	$865,269	$3,350,830	74.2%
Marquette	339,113	1,088	50%	32%	17%	$58,221	$5.04	$1,708,953	$907,954	$2,616,907	65.3%
Traverse	257,449	1,985	61%	45%	22%	$60,880	$5.12	$1,317,378	$368,919	$1,686,297	78.1%
Margaret	253,922	1,391	50%	32%	17%	$55,577	$8.33	$2,115,799	$623,060	$2,738,859	77.3%
Bradford	228,054	2,194	69%	56%	42%	$89,328	$9.32	$2,124,589	$686,681	$2,811,270	75.6%

Using the Thompson Stores example, an analysis and prioritization of all markets throughout the United States that have at least 250,000 people would be conducted. For the purposes of this analysis, Primary Metropolitan Statistical Areas (PMSAs) as defined by the U.S. Bureau of the Census, will be used to define the geographic extent of each market. For example, the Detroit market would include the five-county area surrounding Detroit as indicated by the United States Census. Therefore, competitive and demographic information that would be collected would include data for

Oakland, Macomb, Wayne, St. Clair, and Monroe Counties, Michigan for the purposes of analyzing the Detroit market.

Develop Market Groupings by Size

The second step in conducting a market prioritization for Thompson Stores is to group markets by population. This process is important to the analysis because it provides the foundation for an equitable evaluation for markets of like size. For example, major U.S. markets such as Los Angeles and Chicago should be considered in the context of other major markets (e.g., New York and Philadelphia). Further, in very simple terms, large markets (e.g., Los Angeles) will usually appear to be more viable expansion opportunities than small markets (e.g., Tucson); by virtue of their density, large markets will usually be able to support many more stores than smaller markets. However, a small market such as Tucson might be capable of supporting higher volume, more profitable units. For these reasons, it is usually advantageous to group markets by size when assessing their viability.

To establish appropriate population groupings for the purpose of market screening, the process usually begins by determining the population required to support a single store. Therefore, for the Thompson Stores example, the first grouping would include markets with between 250,000 to 450,000 persons; this has already been established to be the minimum size market needed to support a single Thompson Store. Next, a population level for markets that could support between two or three stores would be established. Markets included in this category would likely have a population base that is roughly double the one-store category. Therefore these markets will have a population base greater than 450,000 but less than 1,000,000. Other appropriate market groupings may include:

- 1.0 to 2.0 million population — Markets that could support three to five stores

- 2.0 to 4.0 million population — Markets that could support six to nine stores

- Greater than 4.0 million population — Markets that could support more than nine stores

Obviously, this process will vary by retailer depending on the population required to support a store. For example, the minimum threshold to support a convenience retailer such as a supermarket is between 40,000 and 50,000 persons, whereas destination-oriented retailers may require up to 1,000,000 persons or more to support a store. Clearly, the size (and number) of the various market groupings would be significantly different for these two retail types.

Competitive and Demographic Data Collection Process

Having grouped markets by size, the next step in the prioritization analysis is to collect and process all relevant competitive and demographic data by market. It has been determined via the database development process that the only significant competitors to the Thompson Store concept are large-format book stores. For the purpose of this analysis, an accurate count of all existing and planned large-format book stores within each market will need to be determined. To allow for meaningful comparisons from market to market, competition is usually measured in terms of units per capita or units per household. In this example, it may be appropriate to calculate the ratio of competitive units per 250,000 persons, roughly equivalent to the minimum density needed to support a single Thompson Store. This level of measurement allows the analyst to determine whether the population threshold required to support a competitor's store is comparable to the Thompson Store threshold, and also provides a strong indication of each market's competitive environment.

All other competitor formats were found to have little direct impact on Thompson Stores. However, this does not mean that other, less directly competitive formats should be ignored. Rather, a simple index of the number of such outlets per 250,000 population, for example, will allow the analyst to effectively monitor the indirect competition in new markets.

The most efficient and accurate method of collecting competitive data by market is to contact the competitors directly for a list of their stores. More specifically, a call can be placed to each competitor's corporate public relations department for information regarding all existing and announced store openings, or the analyst could access the Internet home page where many retailers maintain a listing of all existing and planned locations. In the absence of publicly available competitive locations, other sources such as local chambers of commerce, planning departments, Yellow Pages (both CD and paper formats), newspaper ads, directory assistance, or real estate brokers can be used to obtain the necessary information. Once this data is collected, a calculation of competitive units per capita is made to provide an indication of the relative "competitiveness" of one market versus another.

The demographic characteristics evaluated for each market under consideration are a result of the correlation analysis conducted as part of the database development. As mentioned previously, the sales performance of Thompson Stores are most highly correlated with the following variables:

- Percentage of households with incomes greater than $100,000

- Percentage of population employed in executive, professional, or sales occupations

- Percentage of the population with college degrees

These demographic variables by market are typically collected by accessing one of the national demographic data vendor's databases. It is important to note that the prioritization analysis is based on the average characteristics for each market as a whole, and does not take into consideration the variance or distribution of these variables throughout each market. The purpose of prioritization analysis is not to determine specifics regarding the locations of possible site opportunities for a chain within each market; rather, it is meant to provide an indication of the overall viability of one market versus others of comparable size for the specific retail concept being considered.

Development of the Profile Score

The process of producing a profile score for each market is the most important step in the prioritization analysis. It is imperative to the accuracy of the prioritization analysis that the strength of each variable (e.g., demographics, competition) considered during the scoring process reflect its relative importance with respect to store performance. The strength of each variable is typically determined during the correlation analysis which is conducted as part of the database development process. For example, if the income variable (e.g., households with incomes greater than $100,000) has the strongest positive correlation value, then it should have correspon-ding significance when weighting the demographic and competitive variables within each market in the prioritization process. The following example illustrates how we would develop a scoring system for Thompson Stores using correlation values established in our database development.

TABLE 6-2
SCORING SYSTEM DEVELOPMENT

Variable	Correlation Value
% Households with income > $100,000	0.1578
% Executive/professional/sales occupations	0.1542
% College-educated population	0.1506
Direct competition	– 0.0564

Table 6-2 presents the correlation values for Thompson Stores determined during the development of the store database. Having established the values for each variable, they are used to weight the importance of each variable considered in the prioritization analysis. From the above example, we know that each of the demographic variables have roughly equal correlation values and, therefore, should have the same influence on the scoring process for the prioritization analysis. Additionally, we know that the correlation value for competition indicated a weak negative influence, roughly one-third the strength of each of the demographic variables. Therefore, the relative importance of these variables can be reflected in the prioritization analysis by assigning the following weights to each:

TABLE 6-3
SCORING SYSTEM DEVELOPMENT ADDING WEIGHT

Variable	Correlation Value	Weight
% Households with income > $100,000	0.1578	30%
% Executive/professional/sales occupations	0.1542	30%
% College-educated population	0.1506	30%
Direct competition	-0.0564	10%

These weights are used to develop a composite index for each market which can then be used for prioritization purposes. Once the composite scores are completed, the markets can be ranked within

TABLE 6-4
PRIORITIZATION OF COMPOSITE SCORES

Market	% HH with Income >$100K	% Exec/Prof/ Sales Occupations	% College Educated	% Direct Competition	Total Score
(Weighting)	30%	30%	30%	10%	
Washington, D.C.	100	100	100	69	96.0
Los Angeles, Long Beach, California	47	33	87	50	55.1
Chicago, Illinois	30	62.5	67	56	53.5
New York, New York	51	54	0	100	41.5
Philadelphia, Pennsylvania	25	12.5	47	37.5	29.1
Detroit, Michigan	0	0	13	0	3.9

each density grouping from high to low to identify which markets exhibit the most favorable profile based on the variables considered in the prioritization analysis. In the hypothetical example, composite scores for the density grouping of markets with greater than 4,000,000 persons would be calculated as presented in Table 6-4.

The above example indicates that the Washington, D.C. market would be the most attractive large market for Thompson Stores, with Los Angeles and Chicago offering secondary options. The balance of the markets considered in this group represent less attractive options for expansion of the concept. Based on this information, Thompson Stores management now has the requisite data from which to make an informed decision as to which markets offer the greatest likelihood for success.

The findings derived from this analysis would be considered in conjunction with economic factors (unemployment rates, land costs, etc.), logistical considerations, and operating and marketing issues to arrive at a selection of markets which provide the most favorable overall opportunities for future expansion. Once the market selection process has been completed, the next step is to identify trade areas and corresponding sites within each market that will provide the most effective coverage and sales penetration of the selected markets.

7 The Market Strategy: Selecting Sites in New Markets

A RETAILER WHO IS CONTEMPLATING EXPANSION INTO new markets usually initiates the process by conducting a screening study. Based on this analysis, and allowing for the retailer's appetite for new stores, the retailer may elect to proceed with a single market, prioritize a strategic combination of markets, or target the most viable region of the country. Whether a single top priority market is selected for further analysis or whether several new markets are identified, the next step in the research procedure is the same—a detailed evaluation to identify the store deployment strategy that optimizes a retailer's position.

Developing a store deployment strategy is a two step process. An initial understanding of the dynamics of the market is provided by an in-office analysis. Based on this analysis, the strategist formulates several preliminary deployment strategies, which are then refined based on field observations.

The second step involves a field evaluation to collect the most current information regarding demographics and competition, to test assumptions implicit in the in-office analyses, and to gain firsthand insight into the dynamics of the market. A final deployment strategy is then formulated.

Preliminary In-Office Market Evaluation

An in-office evaluation uses data and information obtained from secondary sources such as data vendors, competitive lists, and the like. While this information is not completely accurate, it is sufficient to provide a general insight into the market, and to facilitate the formulation of several alternative deployment scenarios which can be refined after fieldwork is conducted.

To become familiar with a market, the analyst generally begins by mapping basic market variables (demographics, competition, or other relevant data) at a suitable scale. These maps may be as simple as transparencies overlaid on commercially available road maps. These can be more sophisticated, using maps generated from interactive GIS software which allow the user to create maps at virtually any scale, and map with virtually all relevant demographic and competitive variables. These maps illustrate each market and the elements that influence a store-deployment strategy—competition, key demographic indices, and local and regional access. Typically, these market "overview" maps are based on data from the Census of Population and Housing, Census of Business, or demographic information vendors. They show information such as population or household distributions (existing and projected), demographic characteristics (typically one to three variables which reflect the profile of core customers), competitive locations, and any other data relevant to the analysis. Each map developed for the analysis should also include major roads and highways, as well as basic physical features (rivers and lakes) in order to incorporate access patterns and physical barriers into the deployment strategy.

It is also useful to map major retail facilities, particularly shopping centers, coding them according to size to indicate their significance. In many instances, shopping center data can be supplemented by information obtainable from local newspaper advertising or research departments that have undertaken analyses of retail facilities and consumer shopping habits in their respective areas.

The focus at this stage of analysis is direct competition. Indirect competition is usually present in all markets to roughly the same degree; therefore, within limitations, it can be treated somewhat as ever-present background "noise."

To illustrate how a preliminary strategic analysis might be developed, a series of maps depicting the Detroit, Michigan market has been prepared (Map 7-1, a–e). This simplified example comprises three demographic overlays on a base map consisting of major roads, highways, and physical features such as topographic features and major bodies of water. The thematic overlays depict the distributions of households Map 7-1(a), median income Map 7-1(b), household ownership Map 7-1(c), and white-collar occupations Map 7-1(d) throughout the market. In this example, the retailer operates 25,000-square-foot home accessories stores; its best (core) customers own their homes, are employed in white-collar occupations, and have relatively high income levels Map 7-1(e). The researcher initially reviews the demographic map overlays to identify areas in which in-profile consumers are concentrated within the market, while also looking at the household distribution map to ensure that these areas have sufficient densities to support a store. Typically, areas that appear as if they may have adequate concentrations of in-profile consumers to support a store, are designated as potential deployment opportunities on the market overview maps.

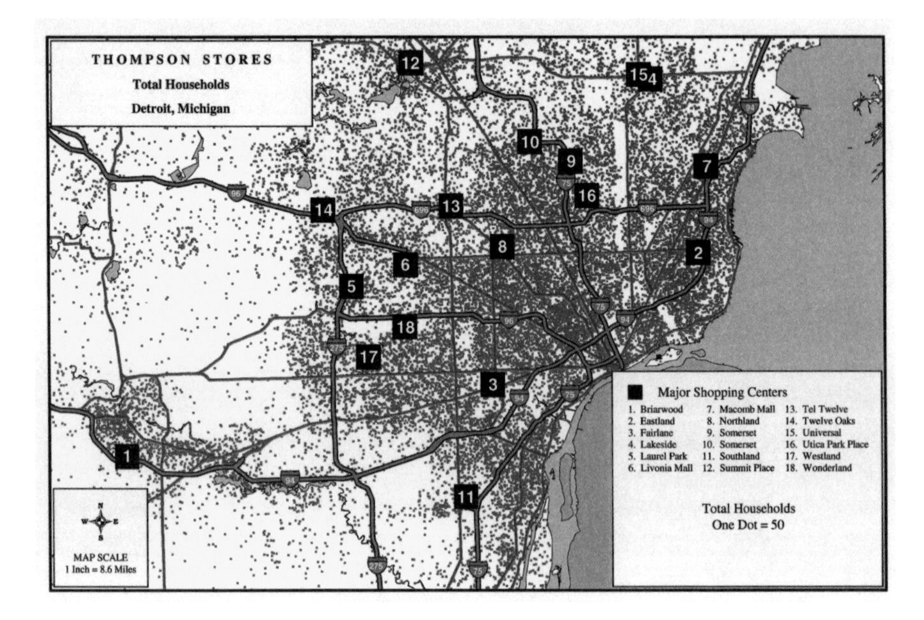

map 7—1a

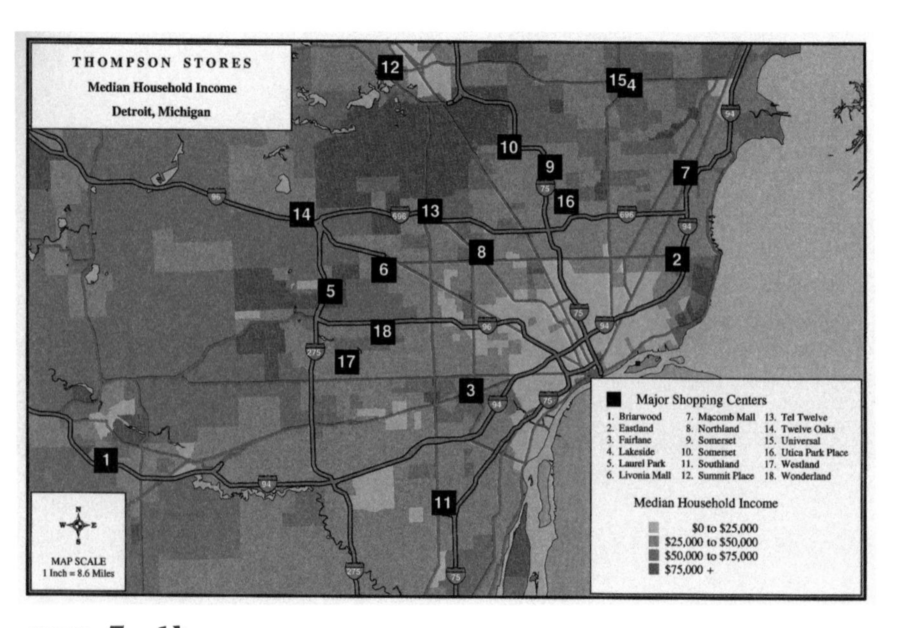

map 7—1b

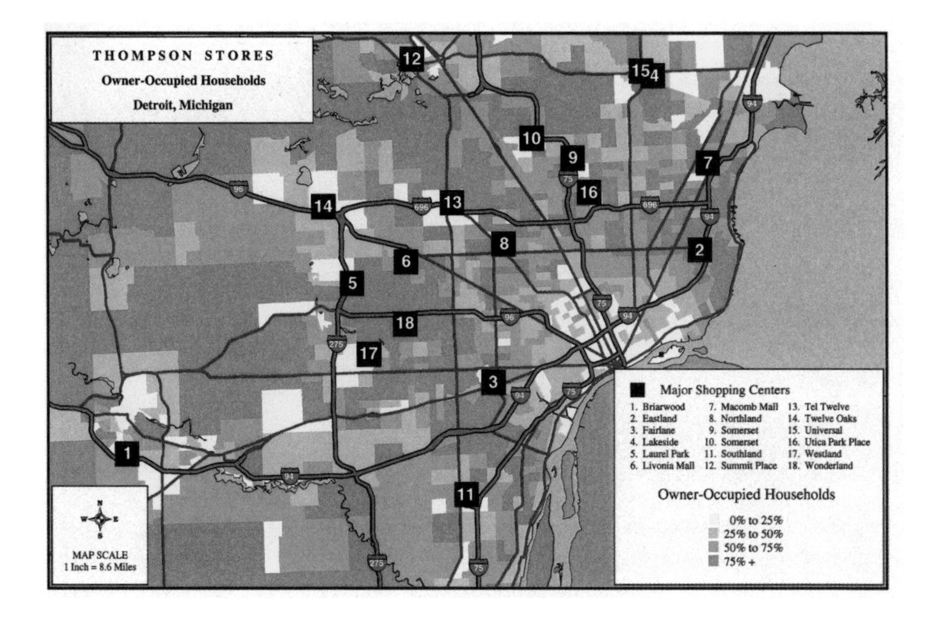

map 7—1c

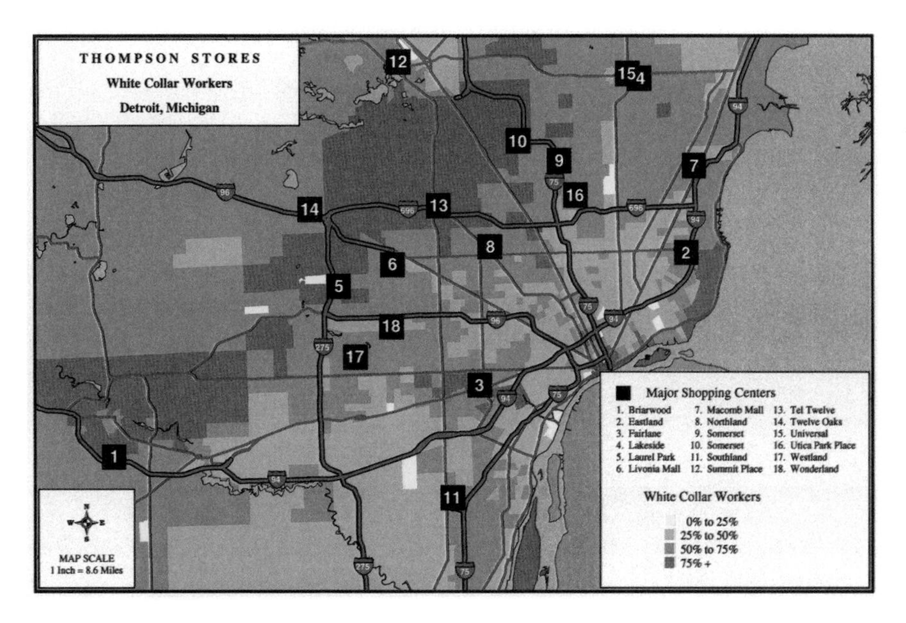

map 7—1d

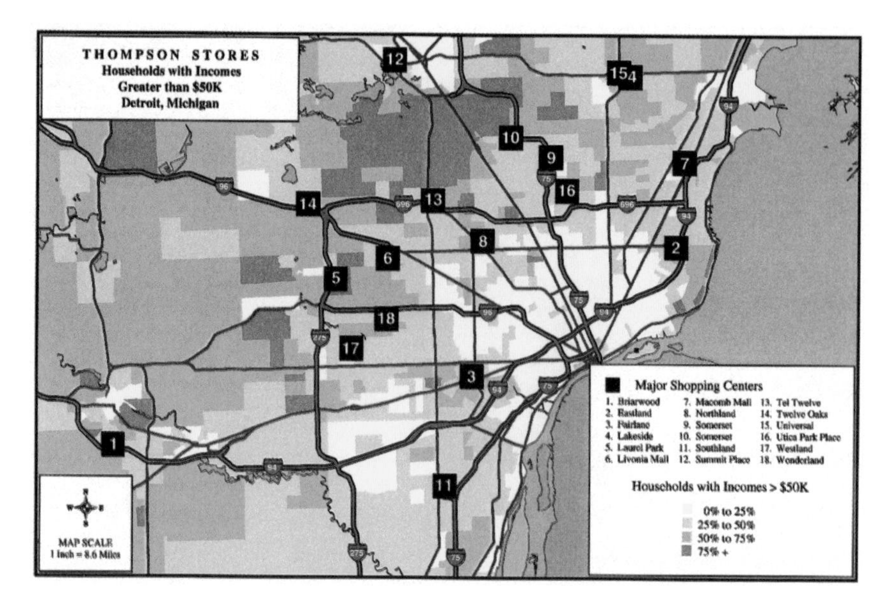

map 7—1e

In addition to maximizing the number of in-profile consumers in the vicinity of potential sites, many retailers also look for sites that offer retail synergy—that is, a concentration of other retailers that either offer complementary goods and services, or provide strong drawing power for the target customer. For example, a home accessory retailer would typically locate its stores in the general vicinity of regional malls or power centers. By placing all the regional malls and power centers in the Detroit metropolitan area on each map overlay, a researcher is able to incorporate the existing retail environment, as well as the demographic and household distributions of the market into the deployment strategy.

The next step in the process of identifying the optimal deployment strategy is to examine the area around each hypothetical "site." A radius is defined around each site, based on the typical reach (or trade area) of the chain. In this example, an eight-mile radius was chosen because it approximates the typical drawing power of the hypothetical home accessories store. As Table 7-1 indicates, there are 20 locations within the Detroit metropolitan area that appear to contain a sufficient number of households within an eight-mile radius to support a home accessories store; obviously, this step is only a "first-pass" with respect to identifying viable site opportunities, as it does not consider factors such as demographic characteristics or competition. Comparing the table with the market overview maps reveals that the highest density trade areas have the least favorable demographic characteristics relative to the concept's customer profile. Conversely, the areas with the smallest overall household densities tend to contain the most favorable in-profile demographic characteristics, and to be the areas with the highest expected growth. Further, based on household density exclusively, Table 7–1 indicates Briarwood Mall would appear to have the least potential for the home accessories concept.

TABLE 7-1
SUMMARY OF HOUSEHOLD DENSITIES WITHIN 8-MILE RADIUS

Location	Households 2001	Households 1996	Households Change 1996-2001
Briarwood	216,780	215,276	0.007
Eastland	730,985	759,620	– 0.038
Fairlane	913,899	958,567	– 0.047
Fairlane North	995,629	1,043,155	– 0.046
Lakeside	450,036	438,192	0.027
Laurel Park	532,420	525,536	0.013
Livonia Mall	743,856	746,331	– 0.003
Macomb Mall	591,953	599,579	– 0.013
Northland	1,042,673	1,066,759	– 0.023
Oakland	775,806	778,463	– 0.003
Somerset	620,971	610,369	0.017
Southland	406,039	422,293	– 0.038
Summit Place	365,243	353,069	0.034
Tel Twelve	698,863	690,590	0.012
Twelve Oaks	3 13,245	293,285	0.068
Universal	945,941	961,807	– 0.016
Utica Park Plac	467,382	452,038	0.034
Westland	578,142	586,836	– 0.015
Wonderland	794,035	811,334	– 0.021

To refine this analysis further, the researcher needs to consider the demographic composition associated with the area surrounding each site. While there is no convenient census statistic that measures the number of target households (i.e., those households which concurrently are nonrenters, affluent, and white-collar workers) that exist in a given area, it is possible to make an estimate. For example, an estimation of the number of target households can be calculated by simply multiplying the proportion of each demographic characteristic by the current household estimate, adding them together, and calculating a simple average of the result. This process essentially gives equal weight to the income, employment, and home ownership levels. The resulting Table 7-2 provides additional

TABLE 7-2
8-MILE-RADIUS CORE DEMOGRAPHICS

Location	Yr. X House-holds	% HH with >$50K	Owner Occupied	White Collar	Target HH
Briarwood	77,569	39.7	47.8	73.6	41,655
Eastland	278,969	28.3	69.2	55.0	141,809
Fairlane	350,467	26.1	65.5	51.5	167,173
Fairlane North	392,528	26.0	64.6	52.1	186,712
Lakeside	162,957	48.9	76.3	63.0	102,228
Laural Park	199,530	48.6	73.9	66.1	125,438
Livonia Mall	283,059	42.7	70.1	65.2	167,948
Macomb Mall	228,318	38.8	77.0	59.0	133,033
Northland	410,959	30.4	64.5	60.2	212,466
Oakland	297,379	42.3	73.8	64.8	179,319
Somerset	237,291	48.5	71.6	71.4	151,471
Southland	161,718	34.7	73.1	51.7	85,980
Summit Place	120,368	48.5	73.3	65.8	75,270
Tel Twelve	279,385	46.8	70.8	71.6	174,307
Twelve Oaks	109,976	58.8	74.1	75.5	76,396
Universal	361,670	34.5	70.1	60.3	198,798
Utica Park Place	166,180	49.5	76.6	63.5	105,026
Westland	218,868	42.4	75.1	60.0	129,497
Wonderland	299,820	38.0	71.6	59.1	168,599

information for each potential location, rank ordered according to the number of target in-profile households. Comparing this new table with Table 7-1 results in somewhat different conclusions concerning the relative merit of each potential deployment opportunity.

With the above data in place, the researcher can now begin a logical analysis of the viability of potential deployment opportunities for the home accessories chain in the Detroit market. Assume that during its database development process, this retailer established that the minimum number of target households required to support a store was 100,000; in this context, the hypothetical sites near Southland Mall, Twelve Oaks, Summit Place, and Briarwood Mall would be eliminated from further consideration.

After eliminating sites with insufficient critical mass, the market analysis now considers other factors that can influence the relative sales potential of each location. For example, in selecting appropriate locations to serve Oakland County, the researcher must weigh the strengths and weaknesses associated with each target location and its corresponding trade area composition. More specifically, the southern portion of the county could be served by either a Northland Mall or a Tel-Twelve Mall location. While Northland has the largest number of target households within an eight-mile radius, it is perceived to have a higher crime rate, and would not provide the high-end retail synergy desired. Even though Tel-Twelve Mall has 38,000 fewer target households, it may represent a superior alternative to Northland as it is perceived to be a safer shopping environment.

As the previous example demonstrates, considering the density and demographics associated with hypothetical sites, in conjunction with more qualitative issues (i.e., safety, appearance of anchors, and so on) allows for the identification of one or more deployment scenarios to serve the market. The ultimate goal is to determine the combination of locations which best serve the in-profile consumers residing throughout the market, while providing complete market coverage. For example, the researcher may have identified the following as alternative store deployment scenarios for the northern portion of the Detroit market:

Scenario 1—Northland, Oakland, Utica Park Place, Eastland
Scenario 2—Tel-Twelve, Oakland, Utica Park Place, Eastland
Scenario 3—Northland, Somerset, Utica Park Place. Eastland
Scenario 4—Northland, Somerset, Utica Park Place, Eastland, Macomb

This type of analysis is conducted throughout the entire market to produce a set of alternative store deployment scenarios for the market as a whole. Furthermore, in addition to identifying the most logical opportunities for store deployment, this analysis would reveal portions of the market that would be poorly served by the various strategies under consideration. For example, considering the various deployment strategies identified for the Detroit market, a review of the demographic maps may reveal a concentration of in-profile consumers that has been overlooked by the various strategies because there is not a regional mall or power center that adequately serves the area. These areas represent potential "holes" in the market deployment strategies, and warrant inclusion if complete market coverage is desired. More specifically, under deployment Scenario 1 as outlined above, there appears to be a "hole" in the Bloomfield Hills area in the northwest part of the market, where income and home-ownership levels are very high.

Once the most logical deployment strategy options have been identified, a more rigorous analysis of the relative potential of each deployment strategy is undertaken. Up to this point, a simple eight-mile radius has been used for each site's trade area; for the purpose of preliminarily considering various hypothetical deployment options, these "generalized" eight-mile trade areas are adequate. However, to determine which one or two scenarios would best serve the market, trade areas must be more precisely defined. In defining trade areas, factors such as the distribution of key demographics, access patterns, competition, and the potential locations of other "sister" stores within the strategy should be considered, so that the trade areas for potential sites reflect those of database stores. When the trade area definitions are complete, the demographics associated with each trade area are compiled. This information provides the researcher with a more realistic basis for evaluating the sales potential of each location in the specific strategy. Finally, using one or more of the sales forecasting tools described in the chapters which follow, sales projections are completed for each hypothetical site included within the deployment strategy. This process should be repeated for all alternative deployment strategies. A comparison of the sales forecasts associated with each strategy will reveal the strategy that concurrently would result in the greatest per-store performance and the most complete market coverage.

During the forecasting phase of this process, the researcher may encounter a situation in which two mutually exclusive locations would serve the same trade area. An example in Detroit would be a hypothetical site near Fairlane Town Center (a regional mall) or near Fairlane North (a power center), both of which are located in the southwestern section of the market. The researcher must determine which of these sites holds the greatest sales potential and is most complementary to the other sites which comprise the store deployment strategy. In order to do this, the researcher will need to forecast the sales potential of each site and assess their respective impacts on the surrounding sister stores that comprise the recommended store deployment.

The ultimate objective of a market strategy is to determine the store deployment that captures the greatest sales and profit potential while providing complete market coverage. In most markets, the researcher would probably be able to identify a few "home run" sites. Typically, these are locations that are central to large concentrations of in-profile customers and weak competition. It is not difficult to "cherry-pick" two or three such sites in a market; however, the researcher is encouraged to make certain that these "home run" locations do not actually represent a concentration of in-profile consumers that is so large, that it would be inefficiently served by only one store, thereby resulting in long-term market weakness. Such strategies undermine a chain's overall market share as implicitly, an underdeveloped deployment strategy results in portions of the market that are poorly served. Further, such strategies are vulnerable to competitive encroachment; rather than serving all the sales potential with a location central to the potential, competitors often choose to open two stores on either side of a cherry-picked location, thus significantly limiting the draw of the cherry-picked site.

Thus, by "cherry picking," the retailer may unnecessarily limit the number of potential stores that can serve the market, and leave itself vulnerable to competitive encroachment.

Typically, forecasts are conducted for each deployment scenario assuming that all proposed stores are open. In other words, the analyst assesses the viability of each alternative deployment strategy by evaluating a retailer's position in a market as it would be at maturity with complete market coverage (i.e., with all sister stores open). In reality, a retailer will rarely simultaneously open all of its stores within a market. Thus, for the purpose of establishing sales goals, determining appropriate inventory and staffing levels, it is often necessary to conduct sales forecasts for stores in the context of the chain's anticipated initial foray into a market, and then conduct forecasts for subsequent stores according to their anticipated opening dates.

The market strategy evaluations described to this point are typically conducted "in-office"; there is no field review at this stage, although market knowledge is clearly an important component of this analysis. Moreover, these deployment alternatives and sales projections derived from these evaluations are not meant to be final recommendations. Rather, their purpose is to give direction for the more detailed analysis that is to follow and, in particular, to provide direction as to which portions of the market should be the focus of field analysis.

Field Evaluation

Conducting fieldwork in a new market provides an analyst with an opportunity to gain a firsthand understanding of a market, and to collect the information needed to enhance and fine-tune in-office research. Refining the initial strategies, and developing a final deployment strategy depends on obtaining the most relevant and up-to-date information available; the stakes are too high to chance any omission or error that may result from developing a deployment strategy based purely on an in-office perspective.

Among the tasks to be undertaken by an analyst during fieldwork in a new market are:

- Identifying, visiting and evaluating all relevant competition, including any store not identified from an in-office perspective.

- Driving the market and evaluating traffic and shopping patterns; observing and confirming the nature of access routes–the number of lanes, traffic flow, speed limits, and possible barriers to access.

- Evaluating specific potential sites or locations.

- Evaluating existing retail concentrations and shopping centers, reviewing their tenant mix, physical plants, and apparent vitality.

- Evaluating neighborhoods relative to recommended sites and competitive locations to ensure that their appearance is consistent with the image implied by the demographic information.

- Speaking with local agencies to obtain additional information and updates on data including:

    - population and household estimates and projections which can be obtained from planning agencies and councils of government;

    - information regarding proposed retail developments, especially the location of new or proposed competition; typically this information can be obtained from planning agencies, newspaper advertising departments, or chambers of commerce;

    - planned new roads or road improvements which can typically be obtained from a five-year plan obtained from the public works and the state highway department.

Competitive information derived from retail directories or telephone books is very useful in the formulation of preliminary in-office deployment strategies, but is usually incomplete and does not reflect known future competitive development. For these reasons, competitive data available via secondary sources is inadequate when formulating a final deployment strategy. To ensure that the final deployment strategy accurately reflects the competitive environment which will ultimately be encountered, it is necessary that the analyst visit all of the major competing stores to observe their operations, physical condition, and relationship to nearby and adjacent retail and other developments, including the surrounding neighborhood served by the store. The analyst needs to review the size and scope of any recent or forthcoming competitive changes including the addition of new stores, store closings, expansions of existing facilities, remodeling—in short, anything that may impact potential sales or market share. Fieldwork provides the opportunity for the researcher to evaluate each competitor's operational characteristics such as service levels, pricing, selection, age and condition of store, and site characteristics (i.e., parking, visibility, ingress/egress). There is no substitute for fieldwork, and certainly no better way to confirm the accuracy of the data that may have been obtained in advance of the field review. Ultimately, the researcher will consider these factors when assessing the impact that competition will have on the performance of proposed sites.

The overall shopping patterns within a market are usually strongly influenced by the concentration of major retail developments within a market. Therefore, during field work, the researcher should become familiar with the locations of shopping centers and malls, and their respective anchor tenants, so that their influence on shopping patterns may be incorporated into the deployment strategy. Typically, the researcher will visit the most significant of these retail concentrations to conduct an assessment of their appearance, the strength of their tenant mix, and the number of vacancies; these factors provide clues as to their influence on existing shopping patterns. It is also essential to know the location, size, proposed anchors, and development schedule for any future shopping centers. The type of shopping centers that the analyst will be interested in assessing is typically a function of the type of retail operation that they are deploying in the market. For example, if a deployment strategy is being developed for a convenience retailer (supermarket or drugstore), an assessment of regional malls may not be necessary, whereas if the strategy is being conducted on behalf of a larger format retailer that depends on the synergy associated with proximity to a major retail concentration, then information regarding the location and nature of regional malls would be critical.

For the purpose of an in-office analysis, street maps provide an adequate indication of traffic patterns and major access routes. However, the traffic pattern inferences derived from street maps need to be confirmed during a field evaluation, as factors which are not obvious from maps (e.g., one-way streets, median breaks, traffic controls, physical/cultural barriers, tolls) can significantly influence access patterns. Further, access problems due to an inadequate number of lanes, traffic congestion, road medians, or turn restrictions may impact new store locations or competitor sites. Planned road improvements that will expand traffic capacity or alter existing travel patterns must be checked. Depending on their specifics, they could enhance a site's ability to serve its trade area, be a detriment or have no impact. The analyst should know which of the alternative plans is likely to occur, and to what extent these factors may impact the site's trade area and sales potential.

Driving the market is necessary to enable an analyst to personally experience and understand how access routes and traffic patterns influence shopping habits, and to incorporate these factors into the deployment strategy. Typically, the analyst will want to drive the major access routes and commercial streets (in both directions) during peak and nonpeak traffic periods. Furthermore, a drive-by of potential sites will provide a "consumer's perspective" on site characteristics such as visibility and ingress. The analyst should also periodically drive residential streets within neighborhoods that will be served by the deployment strategy. This allows the analyst not only to gain firsthand appreciation for how each neighborhood fits into the total market, but identifies economic and land-use transition zones that can have a dramatic affect on shopping patterns.

As previously suggested, fieldwork provides an opportunity to visit local planning and other relevant agencies to obtain and review the most current land use, master plan, and population estimates and projections. During these visits, the analyst should confirm the locations of proposed new competition and other changes in the area, including recent and pending zoning changes or challenges related to retail development.

More important than any single fact obtained from fieldwork is the opportunity it provides the researcher to review and verify the various assumptions upon which the initial in-office deployment strategy has been based. Following a field review, the initial deployment strategy should be reassessed, and refinements to the strategy and individual sales forecasts made so that it reflects the incremental knowledge and insight obtained from fieldwork.

Ultimately, the final store deployment strategy will serve as a blueprint for a retailer's store rollout program within the subject market. Such a strategy should clearly indicate the location of optimal sites within the market (from which the real estate department will base its search for actual site opportunities), and provide an estimate as to the sales potential associated with each recommended location. A successful strategy will ensure that all significant concentrations of target demographics are adequately served, thereby minimizing the retailer's exposure to competitive encroachment, and concurrently maximizing overall market share. However, these goals cannot be realized at the expense of acceptable per-store sales performance. As such, a successful strategy strikes a balance between the often conflicting goals of achieving maximum market share and maximum per-store sales performance. A successful strategy will also identify those locations that provide the greatest opportunity for initial deployment, recognizing that as a result of financing and management needs, some phasing in development will be necessary. Future locations for "emerging" areas (areas which currently represent marginal opportunities but are experiencing significant growth) should also be identified, along with a potential timing for their development.

Obviously, a considerable amount of time usually elapses between the completion of a deployment strategy and the opening of new stores. It is important for the researcher to understand this time interval and incorporate it into the deployment strategy. A time frame of one to two years from the date of undertaking the analysis to the day the first store opens is not uncommon; therefore, to the extent that it is practically possible, the deployment strategy should anticipate and reflect the conditions that will exist within the market at the time stores will open.

Finally, when phasing store openings over a significant period of time, it is important that the deployment strategy remain current. After the initial store openings, it is worthwhile to review the market strategy incorporating the sales performance of the first set of stores and assessing any competitive changes that may have developed. If the assumptions inherent in the original development plan did not materialize, a revision of the sales forecasts for stores yet to be built, or a revision of the overall deployment strategy would likely be appropriate.

Every plausible alternative deployment strategy can be evaluated and every store's potential sales and contribution to an overall market strategy can be determined. Through a process that tests alternatives, combined with the requisite fieldwork, the optimal deployment strategy can be identified, thereby significantly enhancing the likelihood of a successful retail store expansion program in new or existing markets.

8 Forecasting Tools: Analog and Regression Systems

IN 1932, WHILE WORKING FOR THE KROGER CO., WILLIAM Applebaum developed the analog sales forecasting technique[1]. Since then, this forecasting methodology has been enhanced considerably by the use of statistical techniques, and computers which have increased the usefulness and accuracy of analogs, particularly for retailers with a strong customer segmentation.

Analogs provide a tabular "snapshot" of a retail store's performance. A series of these "snapshots" provides a historical record of the relationships that exist between a retailer's geographical sales distributions and the demographic, competitive, and site conditions that exist within each store's trade area. The forecasting methodology that results provides the analyst with a means of extrapolating proposed site sales forecasts from the historical evidence of similar existing stores.

In the past, the factors considered in the analog method were largely limited to the distance consumers must travel from the forecasted area to the site, competition, and population (Table 8-1).

The early analog approach worked particularly well with convenience goods stores. Analogs have been embellished greatly in

TABLE 8-1
SIMPLE ANALOG

Zip Code	Zip Name	Driving Distance	1997 Population	Capture Rate	Per Capita Sales
48141	Inkster	1.2	28,839	14.7%	$15.29
48124	Dearborn	2.2	27,224	20.5%	$22.59
48125	Dearborn Heights	2.6	20,862	5.5%	$7.91
48128	Dearborn	2.8	18,728	22.5%	$36.04
48127	Dearborn Heights	4.3	30,274	5.7%	$5.65
48135	Garden City	4.5	27,239	3.5%	$3.85
48184	Wayne	5.0	19,303	2.5%	$3.89

recent years. The new enhanced analogs not only incorporate distance and population density which traditionally have been displayed in analog tables, but also considerable detail about customer characteristics. For example, enhanced analogs illustrate the strong positive correlation between educational attainment and bookstore sales. Underlying the modern analog table is a considerable evaluation of in-profile customer data to sharpen and focus analog location analyses (Table 8-2).

TABLE 8-2
COMPLETE ANALOG

Zip Code	Zip Name	Driving Distance	1997 Population	Median Household Income	% College Graduate	Capture Rate %	Sales	Per Capita Sales
48141	Inkster	1.2	28,839	$27,905	0.089	14.7%	$441,000	$15.29
48124	Dearborn	2.2	27,224	$39,042	0.196	20.5%	$615,000	$22.59
48125	Dearborn Heights	2.6	20,862	$35,024	0.065	5.5%	$165,000	$7.91
48128	Dearborn	2.8	18,728	$44,350	0.281	22.5%	$675,000	$36.04
48127	Dearborn Heights	4.3	30,274	$41,965	0.147	5.7%	$171,000	$5.65
48135	Garden City	4.5	27,239	$37,241	0.092	3.5%	$105,000	$3.85
48184	Wayne	5.0	19,303	$33,002	0.079	2.5%	$75,000	$3.89
	Trade Area Totals:		172,469	$36,823	0.133	74.9%	$2,247,000	$13.03
	Sales from Beyond the Trade Area:					25.1%	$753,000	
	Grand Total:					100%	$3,000,000	

Typically, the demographic and competitive characteristics that have been added to the modern analog table are determined using regression and correlation analysis (discussed in detail below). Since distance is a powerful determinant of sales potential, particularly for convenience goods, regression is used to eliminate its impact on sales, to gain a clearer focus on demographic and competitive impacts on sales. By correlating demographic and competitor data with the sales levels not explained by distance or population (the residual from the regression procedure), a clear pattern of who patronizes the store and the impact of specific competitors can emerge.

Macro- vs. Micro-Analogs

With the analog approach, two alternative submethodologies are possible: macro-analogs and micro-analogs. Macro-analogs generally rely on a careful analytical judgment concerning the demographic and market factors of an entire trade area; they ignore how consumers, competitors, and demographics are distributed throughout a trade area. In what is almost a single-step procedure, an analyst determines that a particular analog store or set of analog stores apply to the site being evaluated. With some minor adjustment, the sales volume for the proposed site is summarily estimated. The sales estimate is essentially the same as that of the analogs being used. For example, if existing stores have annual sales volumes of $18, $20, and $22 million, and their trade area populations, incomes, and competition are the same as the trade area for a potential site, the sales estimate for the new store will likely be in the neighborhood of $20 million. Should there be a strong competitor affecting the proposed site, a downward adjustment of the sales forecast might be made. While this approach provides a quick estimate, it is grossly unresponsive to the systematic evaluation of variations with respect to customer and demographic distributions within the market, and is therefore likely to miss important details.

Among factors that the analyst would likely consider in selecting macro-analogs are the region of the nation in which the proposed site is located, advertising expenditures, competitors, cost of living, population density (including density by type of household—renter, owner, single family, multiple family), demographic components of the trade area, and scale of the market (single-store opportunity versus multiple stores). Situational characteristics such as access, synergy with adjacent land uses, parking, and positive and negative impacts of competitive clustering may also be considered when selecting macro-analog matches.

Today, many retailers still rely on macro-trade area analyses. Tables 8-3 and 8-4 (on pages 106 and 107) show data that, from a mathematical point of view, are precisely the same although the location implications for developing a new store are entirely different. The household income, housing values, proportion of single-family home owners, and age statistics show no difference, either in mean values, modes, or medians—until location and distribution over geography is considered.

TABLE 8-3
Micro- vs. Macro- Perspective on Trade Area

Zip Code	Driving Distance	1997 Population	Median Household Income	Housing Value (000's)	Single Family Homeowners	Median Age
10001	0.5	2,720	$16,457	$45.5	45.6%	28.1
10002	1.3	3,121	$16,892	$56.8	50.2%	29.5
10003	1.5	4,450	$24,583	$55.4	60.2%	34.5
10004	1.7	9,537	$28,456	$67.5	61.4%	35.1
10005	2.3	12,457	$34,572	$85.6	75.2%	37.5
10006	2.5	15,706	$52,618	$150.5	80.1%	42.5
10007	2.9	18,279	$48,852	$146.6	82.2%	38.5
Trade Area:		**66,270**	**$39,661**	**$110.2**	**72.9%**	**37.7**

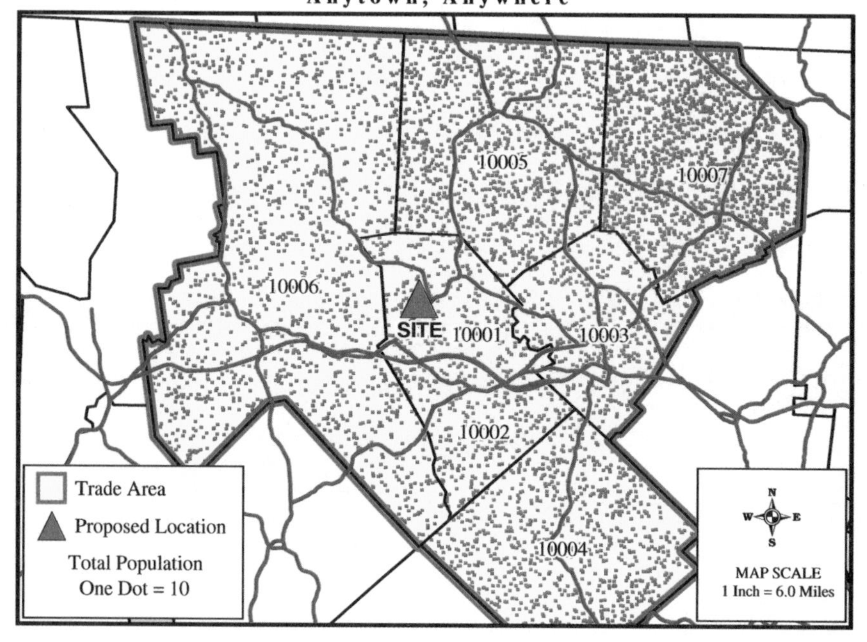

map 8—1

TABLE 8-4
Micro vs. Macro Perspective on Trade Area

Zip Code	Driving Distance	1997 Population	Median Household Income	Housing Value (000's)	Single Family Homeowners	Median Age
20001	0.2	9,842	$60,875	$185.6	95.5%	42.5
20002	1.2	10,420	$62,410	$175.5	84.5%	44.2
20003	1.4	16,524	$47,852	$122.4	72.9%	37.5
20004	1.7	9,537	$28,456	$67.5	61.4%	35.1
20005	2.1	8,125	$16,984	$55.7	60.1%	33.5
20006	2.2	5,541	$14,250	$42.3	58.8%	33.5
20007	3.1	6,281	$15,892	$46.5	48.9%	32.5
Trade Area:		**66,270**	**$39,661**	**$110.2**	**72.9%**	**37.7**

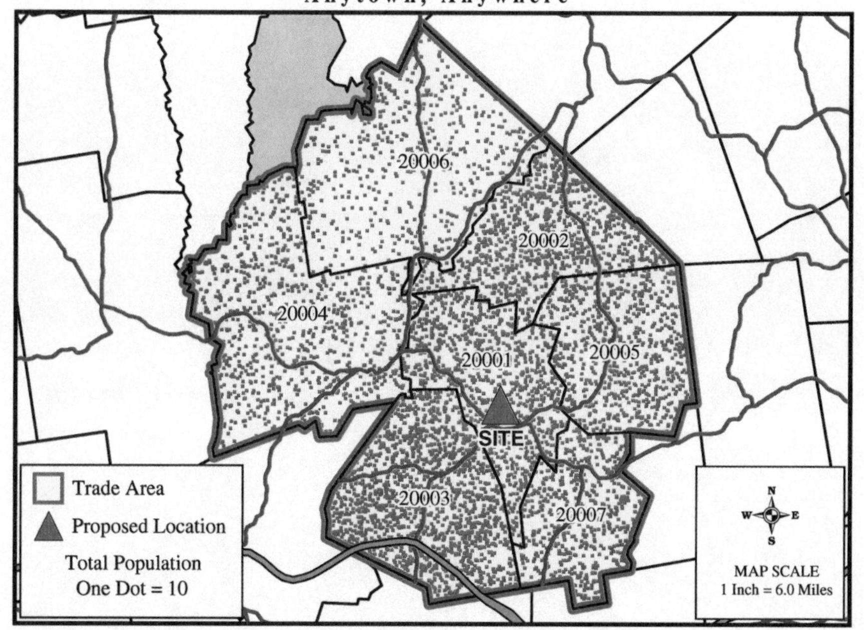

map 8—2

Clearly, the macro-analog approach to the overall numerical data in this market can be misleading. In contrast, the micro-analog approach is able to derive the sales potential value from the different spatial distributions of both basic and subtle market variables. As a result, higher levels of forecast accuracy are generally achievable using the micro- (or disaggregate) trade area analog approach as it more satisfactorily handles the demographic or competitive variations within the trade area under consideration. While the macro-approach may serve to provide a rough sales forecast in the initial stages of an analysis, this forecast should be challenged and changed as component disaggregate (micro) trade area numbers are developed.

Micro-analogs break the trade area into its component geographic parts, typically a standard geography such as block groups, census tracts, or ZIP Codes. The significant advantage associated with this strategy is the ability to differentiate patterns of sales within the trade area. For example, if the eastern half of the trade area is characterized by high levels of direct competition and out-of-profile demographics, very low levels of sales would be expected. If the western half of the trade area is only lightly competitive and has in-profile demographics, very high levels of sales would be expected. In total, however, the hypothetical trade area is average in all respects. If this average were to be viewed only as a "macro"-analog, it would hide both the reality of the sales pattern within the trade area and any useful insights about the relationships between demographics, competition, and sales that can be applied to a new forecasting situation.

Constructing an Analog Database

A significant difference between the analog method and the other sales forecasting methodologies is its adaptability to assess virtually all types of retail stores. In contrast, the gravity model (discussed in detail in Chapter 9) is used primarily in the supermarket and drug store industries, whereas regression analysis is used primarily for specialty stores which have an appeal to specific "in-profile" groups of consumers. The analog method works well not only for the widest variety of store types but also has the longest proven record of successful, pragmatic results.

The modern analog database is derived from customer, competitor and demographic data gathered for a series of representative stores. Next, sales are estimated for each sub-geography surrounding the store. Typically, these sales are expressed on an annual or weekly basis as determined by what is most appropriate and/or widely accepted for the store type in question. Capture rates are then determined by dividing sales for each sub-geography by the total sales for the store. Trade areas may then be defined by following the procedures outlined in Chapter 3.

A forecasting database, however, takes the process one further step by classifying the entire set of analogs by trade area type. Few, if any, modern retailers operate in only one type of market or serve only one type of trade area. Typically, a retailer will serve both large and small metropolitan areas, as well as trade areas that are inner city, suburban, and exurban, to name a few. An effective forecasting database recognizes that there are many such situational aspects of retail trade areas by retail firms. The reason for this attention to the number and type of stores to be included in a forecasting database is to ensure a sufficiently robust subsample of analog stores for every foreseeable type of location to be forecasted in the future. This classification process ensures that like trade areas are defined in a like manner. As a practical matter, this is a rather straightforward process that classifies each trade area into a group of similar trade areas. Each group should represent a consistent pattern both in terms of its geography and overall trade area capture rate. In other words, if suburban trade areas typically extend inbound six miles toward the inner city and outbound ten miles toward the rural fringe, and capture 75 percent to 80 percent of total store sales, then a like pattern would be expected for all similarly situated store locations. Likewise, if trade areas typically extend farther along a limited access freeway, that pattern would also be expected to be regularly repeated.

As always, however, the actual application of the classification system is never quite as smooth as the theory suggests. A problem commonly encountered when creating subcategories within an analog set is accounting for significantly underperforming stores. Poor store performance is often reflected in the geographic extent of a store's trade area and can lead to some apparent inconsistency within the store groupings. In other words, if one store in a group has an unusual geographic pattern in its trade area, it is usually an indication of an underlying performance or competitive problem. An even more common problem is the process of identifying boundaries to the classification scheme. For example, where does "urban" stop and "suburban" begin? While there are no clear-cut answers to these problems, it is vitally important to go through the process for two reasons. First, it is essential in communicating how to define trade areas for proposed sites for which a forecast is to be conducted. Typically, several different individuals will be called upon to use the analog forecasting database over a period of years. Without consistency of trade area definition, any ensuing forecast may be severely compromised. Secondly, without consistent trade area definitions, any statistical analysis will be degraded and, in severe cases, such analysis may be entirely unable to uncover the underlying variable relationships.

After trade areas have been defined and made consistent, correlation and regression techniques are employed to determine relationships between the predictor variables (usually competition and demographics) and the dependent variable (usually sales by geographic subarea). This process is similar to that discussed in Chapters 4 and 5. Based on the results of this analysis, the final disaggregate analog set can be assembled.

The analog set might typically be organized by region of the country, market type (e.g., major metropolitan area, rural market), and location type (e.g., urban, suburban) to better illustrate the similarity and consistencies within subgroups. A similar process of organization would be undertaken for "macro"-analogs (see Table 8-5). While the overall process remains the same, trade areas defined for macro-analogs are often derived from mile rings, or other very simply defined trade areas.

TABLE 8-5
MACRO-STORE ANALOGS

	% Single Family Owner	% Single Family	% Owner	Median HH Income	% Inc. >$70K	% White	Pop. in Trade Area
Lancaster	63%	70%	67%	$41	32%	88%	288,743
Grand Island	59%	65%	64%	$39	32%	85%	203,475
Schenectady	48%	55%	52%	$40	33%	88%	211,657
Randallstown	31%	54%	66%	$43	35%	50%	143,697
Rockville	48%	69%	74%	$69	50%	77%	155,655
Sandy Spring	51%	65%	68%	$60	49%	70%	189,765
Springfield	50%	57%	55%	$37	30%	84%	233,454
Brighton	32%	46%	36%	$49	37%	69%	143,698
Randolph	36%	47%	40%	$42	36%	69%	155,787
Levittown	61%	72%	74%	$53	45%	93%	123,454
Bangor	52%	69%	75%	$40	33%	93%	132,453
Malvem	50%	67%	72%	$55	43%	95%	169,450
Avg. for Trade Areas:	50%	62%	57%	$41	37%	81%	179,274

Macro-analogs sacrifice much of the geographic detail found in disaggregate analogs which provide the analyst with a direct perspective on how demographics and competition impact sales performance.

The "micro"-analog forecasting approach parallels that of map sectors in a gravity model. A major difference is that the micro-analog is usually applied to stores with geographically large trade areas; accordingly, the components of the disaggregate trade are (DTA) are larger than the block groups characteristic of supermarket or drug store trade area. Nonetheless, the micro-analog method can work equally as well with disaggregated spaces as small as blocks or block groups. An example of a disaggregated sales forecast is presented in the table below showing each component of the trade area of a proposed store and the estimated sales potential evaluated for each market subsegment.

Using a micro-analog forecasting approach, each sector (or sub-geography) in the proposed store's primary trade area is individually forecasted and the forecasts for each sector are added to obtain the projected sales volume originating from within the defined primary trade area, generally about 75 percent of total sales. Sales originating from consumers residing beyond the primary trade area are estimated using macro-analog techniques. In instances, where there are well developed sales patterns in areas peripheral to the primary trade area, projected sales from beyond the primary trade area may be divided into a secondary and tertiary trade areas.

Applying a micro-analog database for the purposes of forecasting sales for a proposed store involves two basic steps: (1) identifying stores that are analogous to the forecasting situation at hand on a macro basis, and (2) finding the most analogous disaggregate matches between the subset of the database and the respective trade area segments to be forecasted. Matching analog stores with the store to be forecasted is both an analytical and judgmental process similar to that for macro-analogs described earlier. Identifying the most comparable analog stores begins with analysis of the geographic extent of each analog store's trade area, its population density, and the geographic considerations of the trade area such as barriers, terrain, and so on. From the regression/correlation analysis process, the analyst has identified key forecast variables that are closely related to sales performance (e.g., demographic and competitive considerations). Such customer profile demographic variables, as well as the competitive environment, would then be considered to ensure that the list of analogous stores have characteristics that most closely reflect those of the proposed store. One caveat should be made: there will never be perfect matches on a macro basis. Rather, it is at this point that the experience and judgement of the analyst most clearly comes into play. There is no substitute for the direct experience of an analyst who has evaluated the database stores directly in the field.

Once store level analog matches have been identified, the detailed disaggregate forecast can begin. That is, a sales forecast is conducted for each segment (e.g., ZIP Code, block group, etc.) within a proposed site's defined trade area. This process involves scanning the subset of analogous stores for those segments that are most comparable with respect to their distance from the store, their demographic composition, and their competitive and situational characteristics. In production forecasting environments, automated analog retrieval programs speed the process of forecasting for disaggregate trade areas. Using an analog retrieval program, the analyst would specify the degree of similarity required for a "match" considering several key factors such as driving distance from the segment to the site, customer demographics, and competitive variables. For each specific retailer, a forecast protocol can be readily established. For example, a fine furniture retailer may first screen the analog database for matches on income, occupational status, and subsequently, driving distance and population density. Conversely, the protocol for a discount store might focus on distance, density, and direct competition, and only secondarily on customer demographics.

For the purposes of forecasting sales for each sector within a proposed store's trade area, the analyst scans for a minimum of four or five analog subareas that are essentially similar on all "match" criteria. A more ideal number would be twelve to fifteen areas that are similar. More than 20 analog matches for a subarea forecast may be indicative of search criteria that is too loose. Having identified the subareas that are the best matches, a geographic review of each candidate analog match takes place to ensure sufficient similarity between the analog match and the area to be forecasted. This review is essential as it will determine how comparable the analog matches are relative to the area being forecasted, with respect to situational issues such as the quality of access, barriers to access, and the relative positioning of the store and competitive alternatives. To make these comparisons, an accurate map of the analog stores' trade areas is required, as well as real-world knowledge of both the analog and forecast store markets. As was reviewed in discussions concerning fieldwork earlier in this text, the well-prepared analyst will have a good working knowledge of each market's current shopping patterns, social and economic geography, and road networks. The bottom line is the ability to make an informed judgment as to the similarity of the analog and forecast situations. This geographic review is further useful in paring the list of potentially analogous situations. In essence, the analyst has pared a list of perhaps as many as 5,000 disaggregate analog examples to the 8 to 12 analog matches that are most comparable to the situation being forecasted.

In using the analog method, there is a choice between "averaging" or "convergence" in estimating the final forecast for each trade area sector. Using the averaging approach, a forecast is developed for each trade area sector based on the subset of analogs that have conditions that are reasonably similar to those in the sector under scrutiny. The analyst could average the sales penetration levels (usually expressed as sales per capita) derived from the comparable analogs and apply this average to the trade area sector being forecasted. The convergence approach focuses on the sub-geography forecast on where the analog penetration levels seem to converge (i.e., what seems to be the prevalent level of penetration level performance). In effect, it is a choice between a mean of the analog penetration levels versus their mode. Figure 8-6 indicates the difference in projected penetration levels that would result from using the "averaging versus convergence" approach in a forecast for a trade area sector. All of the analogs selected have demographic, access, and competition characteristics similar to that of the trade area sector being forecasted.

When the penetration levels achieved by the best analog matches are relatively consistent, simply averaging the penetration levels and applying the average to the trade area sector being forecasted is usually appropriate. However, if there are significant clusters of analog match "outliers" (either high or low), then simply averaging the penetration levels and applying this average to the trade area sector could result in a significant error. Rather, a rigorous review of the analog matches should be conducted in an effort to determine the commonalities associated with the analog outliers (e.g., all of the outliers are located in the same markets, all of the outliers are located adjacent to regional malls, etc.). If the factors common to the outliers are also common to the trade area sector being forecasted, then the mean of the cluster of outliers probably represents the best forecast for the sector. Using this approach, the analyst avoids the trap of routinely forecasting average penetration levels when, in fact, more extreme penetration levels may be called for. Ultimately, there are no hard and fast rules for choosing among the alternatives. As with other forecasting methods, inflexible rules for evaluating the data can result in mathematically correct manipulations, but inaccurate forecasts. The object in store location analysis is always the correct forecast and the astute analyst must know when to relax some of the "rules" in order to provide a logical basis for justifying their forecast.

Figure 8-6
Forecasting Analogs

Store Number	Zip Code	Zip Name	Driving Distance	1997 Population	Median Household Income	% College Graduate	Per Capita Sales
642	25049	Kuntzville	2.6	22,457	$37,458	19.6%	$13.04
647	24781	Zackary	2.9	21,521	$38,485	21.1%	$12.56
545	27845	Devonshire	2.8	18,742	$44,265	23.4%	$14.24
425	22458	Jennifer Crossing	2.7	22,331	$35,845	24.7%	$11.25
645	25011	Davidson	2.9	19,457	$32,451	16.4%	$9.65
425	22456	BitterootPond	2.5	23,451	$29,785	12.4%	$8.42
642	25051	Kuntzville	2.6	22,335	$28,747	9.5%	$8.81
647	24785	Zackary Station	2.7	23,551	$27,369	7.1%	$7.41

Be cautious when using micro-analog sales forecasting. The detail in most analog databases is usually so extensive that it is easy to be lulled into a sense of security about a forecast. Detail is not a substitute for accuracy, however. The analysis must not only be based on accurate data and rational assessment of that data, but it must also pass a series of reality checks as discussed below. Further, the analyst should always remain aware that the analog data are estimates that are only as reliable as the distribution of customers.

After going through the detail of a micro analog forecast, the final sales forecast should be compared with the actual performance of database stores in similar areas and situations. Typically, this check uses macro analogs to determine whether the completed forecasts are plausible and stand up in the face of comparisons with comparable analogs, and to ensure that there are no extreme variations. These comparisons are usually conducted at a variety of levels. The analyst may check to ensure that the average trade area penetration level projected for a proposed site makes sense in the context of the actual average trade area penetration levels achieved by database stores in comparable situations. Questions such as: "Does the projected overall volume make sense in the context of the database?," "Is the estimate of sales originating from beyond the trade area reasonable?," should be asked. If there are such variations in the final numbers, there must be a clear rationale to support the findings. If the analogs selected are logical for the site being evaluated, and an analyst does not stray too far beyond the bounds implied by the analog database, it is difficult to be far from the correct sales forecast. Typically, the final check of the sales forecast for the site will focus on the "reasonableness" of sales generated from beyond the trade area, by comparing the per capita ratios between sales generated from within and beyond the trade areas, or by comparing a plot of per capita sales against overall trade area population. For each retail type, there is a different series of cross-check procedures tailored to the retailers specific requirements. In each case, the process is one of convergence; making sure that the forecast and the "macro"-check indices all point to a similar forecast number.

Skills Needed in Analog Analysis

The analog method is so intuitive and the data manipulations generally so straightforward, that almost anyone will both understand and appreciate the technique. Despite the use of large databases and statistics in providing the foundation upon which analogs are built, analog forecasting systems are the least formula driven forecasting technique and, generally are the most intuitively grasped approach to store location research.

Perhaps the most critical skill for an analyst using the analog method is the ability to look at maps of database stores in the context of the dynamics of the location under evaluation. It takes an intimate understanding of the site and situational characteristics associated with the stores which comprise the analog database in order to be able to identify those that fit a given situation. Very importantly, this means understanding the distribution of households or income patterns within database store trade areas, as contrasted to simply knowing about households and income levels within the trade area under consideration.

Recognizing patterns in distributions means being "literate" in the spatial analysis of markets and data (i.e., the instinctive ability of an analyst to scan and read map distributions and recognize the important components and interactions of their patterns). This ability also includes a recognition of patterns that may be blurred and warrant further investigation including possible statistical testing or elaboration. This ability is comparable to "being numerate" (i.e., having an ability to recognize when analog trade area numbers and proportions, as well as demographic data are sensible, likely, and plausible, and also to recognize when there is a need to question the database for possible flaws and inconsistencies).

Uses of Analog Forecasting

Analog systems are particularly appropriate forecasting tools for shopping goods retailers—those retailers who feature items that appeal to a specific customer profile, such as quality furniture stores, home improvement stores and the like. For example, if the customer profile for a quality furniture chain included households with high incomes and professional occupations, groups of similar target customers can be readily identified using an analog system. In contrast, the sales performance of convenience goods retailers (those retailers which have a broad market appeal such as drugstores and supermarkets) are driven by overall population density and ready access to the store; thus analog forecasting systems do not have any inherent advantage relative to other forecasting approaches for these types of stores. Another significant advantage associated with analog systems is their ability to forecast for situations not directly replicated in the analog database. For example, new competitors can be readily adapted into analog systems. Of course, the limits of this ability to "extend" the database are a result of the analyst's ability and the robustness of the database. For instance, when a chain of variety stores encounters Wal-Mart for the first time, it is unlikely that the potential impact of this competitor could be gauged based on the variety store's existing analog database.

The major weakness of analogs lies in their subjectivity. For this reason, successful analog forecasting systems usually employ the types of macro-validation techniques discussed above to minimize the possibilities for analyst bias. Analog systems do not work well unless applied by an experienced analyst with a thorough familiarity of the database stores. A final limitation in considering analogs as a forecast solution is the relatively high cost of developing a suitable database and the expense of developing and maintaining an experienced analytical staff.

Regression Systems

Regression and correlation for retail location research was first used in the late 1960s. Among the early innovators of these techniques was the John S. Thompson Company, the predecessor of Thompson Associates. In recent years, regression approaches have greatly benefited from advances in statistical software which now make it much easier to gain access to advanced statistical analysis techniques. Once calibrated, regression-based forecasting systems are very simple to implement and do not require highly trained analysts.

In using statistical techniques for location research, it is important that the rules of scientific inquiry and statistics be closely observed. The results of statistical models may be dangerously inaccurate despite an appearance of accuracy and precision. The calibration of workable statistical models requires close adherence to the rules of statistics, a thorough understanding of the retail concept being studied, and a pragmatic understanding of the limits of the techniques employed. Having said this, however, the discussion of regression and correlation techniques below is focused on an overview of the concepts, not a technical discussion of the underlying statistical principles.

Regression Basics

Both correlation and regression can be used to show the statistical relationship between two or more variables: correlation demonstrates a relationship between two or more variables, while regression demonstrates statistical causality. A simple regression relationship can be illustrated by a bivariate plot of points such as those shown on Table 8-7. In this example, sales are related to the number of adults with college or greater educational attainment. Regression procedures fit a line (the regression line) through the points in such a way as to minimize the total distance between the line and each of the points. The regression line is described by its slope (or angle), and the point at which it intercepts the sales axis. The slope is described by a weight, a multiplier for each unit on the education axis.

The forecast interpretation of regression requires the user to accept two critical assumptions. To use the relationship between education and sales for prediction, the analyst must be satisfied that the identified relationship provides both necessary and sufficient explanation. A necessary relationship is one that makes good sense (i.e., there is a logical relationship between higher education and sales). The relationship is sufficient when the explanation or causality relationship is complete. To continue our example the relationship is sufficient because education has been statistically proven to have an influence on sales performance and there are no other significant prediction relationships.

Obviously, one predictor or variable will rarely be a completely sufficient explanation of sales; in reality, sales are never a result of just one cause. In this example, accessibility to the site and availability of alternative sources would almost certainly be significant causal factors. Thus, using the previous example, even though the relationship between education and sales is strong and has a sound theoretical basis, if there are other major factors that theoretically should be part of the explanation of sales, there is still not sufficient explanation. In applied regression analysis, explanatory variables all

TABLE 8-7
COLLEGE EDUCATION AND SALES LEVEL

Frequency Percent Row % Col. %	Sales $0- $99K	Sales $100- $199K	Sales $200- $299K	Sales $300- $399K	Sales $400- $499K	Sales $500- $599K	Sales Above $600K	Total
	11	4	1	0	1	0	0	17
Less than	12.22	4.44	1.11	0.00	1.11	0.00	0.00	18.89
2 Years	64.71	23.53	5,88	0.00	5.88	0.00	0.00	
College	91.67	30.77	4.00	0.00	12.50	0.00	0.00	
	0	7	13	2	0	0	0	22
2-4	0.00	7.78	14.44	2.22	0.00	0.00	0,00	24.44
Years	0.00	31.82	59.09	9.09	0.00	0.00	0.00	
College	0.00	53.85	52.00	9.52	0.00	0.00	0.00	
	1	2	9	12	1	0	0	25
College	1.11	2.22	10,00	13,33	1.11	0.00	0.00	27.78
Degree	4.00	8,00	36.00	48,00	4.00	0.00	0.00	
	8.33	15.38	36.00	57.14	12,50	0,00	0,00	
	0	0	2	7	6	7	4	26
Graduate	0.00	0.00	2.22	7,78	6.67	7.78	4.44	28.89
Level	0.00	0,00	7.69	26.92	23.08	26,92	15.38	
	0.00	0.00	8.00	33.33	75.00	100.00	100.00	
Total	12	13	25	21	8	7	4	90
	13.33	14.44	27.78	23.33	8,89	7.78	4.44	100.00

need to be theoretically relevant, provide a complete explanation of sales and, ideally, do so as simply as possible. Three questions need to be addressed to satisfy the necessary and sufficient criteria required of applied regression procedures. First and most obvious, does some statistical relationship exist between the predictor variable (X) and the factor that is being predicted (Y)—usually sales? In other words, are the two variables correlated with one another? Secondly, how strong is the relationship between X and Y? Finally, can a simple rule be formulated for predicting Y from X, and if so, does the X supply a sufficiently complete prediction of Y?

Multiple regression provides a means to model how a variable to be predicted (Y) varies with a set of independent predictors. Thus, the sales of a retail store from a specific ZIP Code (the Y variable to be predicted) may be related to distance from the store and the population of that ZIP Code. In regression terms, the case by case variability of sales is related to the variability of distance and population using a linear function. Some proportion of the variance in sales is accounted for by the variation of distance and population. This proportion is termed the coefficient of determination, commonly referred to as R^2 (R-Squared). The variance in sales not explained by distance and population is termed the error. If the proportion of sales variance explained by distance and population is high and the error correspondingly low, then one might reasonably determine that these two terms provide a sufficient determination for sales.

Many other factors are considered in addition to distance and population from a store. The specifics of competition, the impact of income, the impact of the image of the store in the market, and the price competitiveness of the retailer are all variables that may help to predict sales performance of a proposed facility. For a regression model to be complete, all theoretical relationships must be evaluated to determine their potential contribution to the explanation of sales. The general process includes: (1) posing the initial questions of what access, customer, competitive, site, or market environment variables impact sales; (2) translating these ideas into variables that can be measured and replicated in each forecasting situation; (3) testing the statistical relationships; (4) translating the statistical results back into the language of the original question; and (5) reaching a conclusion as to whether the initial questions are supported by the statistical analysis.

If the analyst has considered all of the variables that are relevant to the prediction of sales and has unbiased data for each of these variables, a multiple regression equation ought to be able to forecast the sales at any location. An equation that might exemplify the approach is as follows:

$$\textbf{Estimated Sales} = a + b_1 x_1 + b_2 x_2 + b_3 x_3$$

where

a represents the intercept value or constant
x_1 represents distance
x_2 represents population
x_3 represents competition and
b_1, b_2, *and* b_3 represent the multiplicative weighting assigned by regression to each variable.

The value or variable (estimated sales in this case) is called the dependent variable. It is dependent on the information and processing of the other variables and is the value being sought. The other variables (x_1, x_2, x_3), those that predict the value, are called the independent variables. This terminology is logical in that, as in this example, sales are dependent on the other variables (e.g., population and income "cause" sales), and the opposite—that income and population are caused by sales—is irrational. Despite the clear logic of this terminology and the unquestionable theoretical basis for the mathematical processes, their pragmatic handling and the contribution of regression analysis to store location is beset with numerous "traps." As with other forecasting techniques, if the powerful advantages of regression are to be used, they must be employed judiciously by a competent analyst.

Several assumptions necessary for using regression techniques are not always adhered to in real-world circumstances. If these assumptions are violated, there can be adverse effects on the reliability of the findings. For example, linear regression is based on the assumption that the data are best represented by a straight line and that the straight line of the regression analysis is the best predictor. In many cases, even though a curvilinear projection may be more representative of the true relationship, the variables can be transformed to enable linear regression techniques to be used. If the regression cannot be successfully transformed, some potential forecast conditions will not be forecasted reliably.

Secondly, many of the variables that could be of consequence may be impacted by each other and therefore, not truly independent of each other. For example, if high income and high education are both factors in bookstore sales, it is not logical to expect them to contribute independently of each other when income and high education levels are themselves related. Such interrelations in the data are known as multicollinearity. If unaddressed, they can provide mathematically correctly processed data which, because of the failure of the variables to be completely independent, generates misleading results. Finally, it is important that the difference between each variable's actual and predicted values (their errors) be uncorrelated. If such a pattern exists (if the differences are correlated), this generally suggests that an inappropriate form of regression has been used.

The multiple regression equation will not have all points which are perfectly described by the equation. For purposes of discussion, we refer to the graph of the regression results as a line, even though a line is only correct when referring to a one independent variable model; a two-variable model would describe a surface, and higher dimensionality describes increasingly complicated surface geometries (i.e., N-dimensional surfaces). The assumption that the regression line best estimates performance alone ought to suggest caution.

The difference between the position of the points, and their position as predicted by the regression line, results in what are known as residuals or errors from regression. In some instances, particularly when points are very far from the line, the residuals are worthy of detailed investigation. The residuals can be analyzed through a variety of statistical techniques and simple observations of the data. For example, extreme residuals may be the result of conditions that are so novel as not to be repeated elsewhere and, therefore, without usefulness to the predictive equation. Such conditions might typically be found for "flagship" stores where performance may be superior to the remainder of the chain. While mathematical rules may suggest that no portion of the sample be omitted, the object in pragmatic forecasting is the correct sales estimate. It may also be reasonable in practice to look at residuals based on something other than a straight line "best fit" of the cloud of points. Again, as in other techniques, a technically competent analyst who is very familiar with the retailer is imperative for the development of a successful forecasting system.

Assessing Regression for Forecasting

It is necessary to avoid inappropriate predictors despite the mathematical correctness of the process by which they are found. For example, it may be that the best mathematical predictor of supermarket sales, as found by regression analysis, happens to be the concentration of households with four or more cars. However, the occurrence of such households is so infrequent that to base an analysis on that variable is not warranted. Further, in predicting grocery store sales, one would be hard pressed to justify the theoretical relationship to multicar households.

Another concern in the use of regression models for retail sales forecasting, is that the trade area component has to follow a common set of rules. For example, the trade area in an analysis may be two miles in radius, but three miles away a freeway exit exists within 1,000 feet of a proposed site. Failure to control for the scope of a trade area can result in misleading findings.

Regression analysis includes a calculation which indicates the amount of the dependent variable that is "explained" by the independent variable or variables. The mathematics represent the fact that not all variation in the sales data is explained by the regression equation. Another way of interpreting the amount that is explained is to say that the error of making a sales estimate is reduced by employing the relationship determined by the regression analysis. The more variables used that represent true causal relationships between the dependent and independent variables, the more accurate the explanation. However, the level of explanation has to be balanced against how much explanation the dataset is capable of supporting. A very common error is overfitting the regression equation. That is, there may be too many explanatory variables for too few observations. In such a case, although the calculated figure indicating the amount "explained" may be improving, the findings are not really explaining the underlying reality of the marketplace but, rather, mimicking the random variations that occur in any sample dataset.

One of the advantages of regression analysis, in addition to the applications suggested above, is that it can assist in sorting out details of a complicated locational situation, such as different sets of restaurant customers: breakfast, lunch, and dinner patrons. It can also help to sort out the impact of different classes of competition such as direct versus indirect competition.

As in analog and gravity model approaches to forecasting, the regression method has achieved its best success in forecasting disaggregate trade areas. As long as there is no significant causal factor that is missed, the approach by subareas will generally balance errors that may occur in different map segments and the randomness of errors will "wash" in the overall evaluation. Moreover, if one portion of the trade area is misforecasted, the error is confined to just that one portion of the trade area, rather than the trade area as a whole.

Regression analysis provides a solid sales forecast for a store that is based on prototypes in the database. However, regression analyses are less appropriate for forecasting situations that are dissimilar to those in the database such as a new retail format or, possibly, an entry to a market in which the retailer has no experience and customer recognition. While regression techniques would seem to readily allow forecasting beyond the range of its calibration database, in reality this rarely provides a satisfactory solution. Extreme variable data values are rarely simply a linear extension of the relationship found in the calibration dataset. In these matters, the analog approach may be a more appropriate methodology. Regression is most useful for shopping goods retail facilities as the customer profile for these types of retailers can be well-differentiated compared to convenience goods facilities. For a constantly and rapidly evolving store type, it may be preferable to use the analog approach. Perhaps the most significant disadvantage of regression forecasting systems is the difficulty of producing a model that both adheres to statistical requirements and meets the business needs of the retailer.

---

1. See Applebaum, William (ed.), Store Location and Development Studies, Worcester, MA, Clark University, 1962 and Applebaum, William, A Guide to Store Location Research, Reading, MA, Addison-Wesley, 1968.

9 Forecasting Tools: The Gravity Models

GRAVITY MODELS WERE FIRST INSPIRED BY THE physical sciences. Newton's Law of Gravity states that two bodies are attracted to each other in proportion to their mass, and in inverse proportion to the square of the distance between them. Gravity models used in site analysis and sales forecasting have been developed based on this law to predict how consumer expenditures will be attracted to various retail alternatives. As an example of this phenomenon, envision two magnets—a large one and a much smaller one. If both are placed at some distance from each other on a flat plane, and iron filings are scattered between them, the law of gravity suggests that the line separating the two magnetic fields (i.e., the areas within which the filings "gravitate" toward each magnet) will be closer to the smaller magnet than to the larger one. In other words, the larger the magnet, the greater the area within which its influence (or pull) will be felt. If the two magnets were of equal size, we would expect the line separating the two magnetic fields to lie approximately midway between them.

The gravity model approach to store location research and site analysis has a lengthy and well-documented history. Over 70 years ago, William J. Reilly undertook studies of retail concentrations and their apparent gravitational pull among customers. His work culminated in what is now known as "Reilly's Law of Retail Gravitation."[1]

Simply stated, Reilly's Law suggests that the boundary of a community's retail trading area is determined by the population of the urban area, the population of competitive urban areas, and the distances between these communities. Specifically, Reilly's formula for retail gravitation states that retail attraction is directly proportional to the size of the two retail trading centers, and inversely proportional to the square of the distance between the two retail trading centers, expressed as follows:

$$D_{A-B} = \frac{d}{1 + \sqrt{P_B/P_A}}$$

in which:

d = the distance, in miles, on major roads, between two adjacent towns, A and B.

P_A = the population of Town A.

P_B = The population of Town B

D_{A-B} = the edge, or boundary, of Town A's trading area, expressed in miles, toward Town B from the center of Town A.

Reilly's Law attempted to explain the relative pull of one town (or retail trading center) upon an adjacent town (or retail trading center), and to describe the effect of the relative pulls on the area between the two towns (or retail trading centers). The relationship between the two towns was described as a magnetic one in which the strength of the magnetic pull of each town upon the other (and especially upon the area between the two towns), was described as a function of the relative size of the magnets (or towns expressed in terms of their respective populations). There tends to be a positive and direct relationship between a town's population base and the number (or strength) of the retail establishments that exist there. One common measurement of this number or strength is expressed in terms of retail square footage. For example, if Town A has a population base roughly twice that of Town B, it can be reasonably expected that Town A will have about double the number of retail establishments (or about double the amount of retail square footage) as Town B.

Reilly's Law was oriented toward describing how retail trading area boundaries are determined for concentrations of retail establishments existing in neighboring communities of various sizes. But one of its major weaknesses was that it measured only population and distance as factors in defining these trade area boundaries; it did not consider strength of operators, merchandising characteristics, promotional activities, or shopper motivation. Further, it centered on communities and concentrations of retail facilities rather than on individual retail establishments.

During the 1960s, academics experimented with Reilly's Law using real-world data. They were working toward developing a more sophisticated, yet more pragmatic tool to model the shopping behavior of consumers. Their efforts were aimed at eliminating some of the limitations of Reilly's Law, while at the same time developing a gravity model that would enable forecasters to predict individual store performance.

The earliest gravity models relied on the assumption that people shop in one community (or retail facility) or another, based strictly on their distance from the community (or retail facility) and the relative size or mass of the community, often expressed either as the population of the community or the square footage of the retail facility. These early gravity models implicitly assumed that the trade area of one community (or retail facility), could not overlap with the trade area of another. That is, these models assumed that consumers resided within the trade area of one community (or retailer) or another, and would only shop the community (retailer) associated with the trade area in which they resided; we know today that this assumption is not reasonable. In addition, early adaptations of gravity models were not able to incorporate issues such as price, store images, variations in retail formats, or other market considerations that have been found to be important variables in explaining retail store sales performance.

In reality, there is a probability (rather than an exact determination) that a person living at a particular location will shop at one store or another (or one town or another), if similar goods, services and pricing are available at alternative places. This probability is greater for towns or stores that are relatively close, but lower for towns or stores that are further away. This probability exists despite concerns regarding access, travel, or time costs. In other words, the reality of the marketplace reveals that, in any given area, not everyone shops at the closest store; there are many other factors influencing the decision.

During the late 1960s, David Huff was engaged by the Super Valu company (a Hopkins, Minnesota–based grocery wholesaler), to further develop a gravity model to be used in forecasting sales of supermarkets. Provided with a wealth of real-world data by Super Valu, Huff's early efforts were met with mixed success. On one hand, his model was able to predict average store sales performance; on the other, it was not reliable in predicting either abnormally poor or exceptionally good sales performance. His conclusion suggested that there may be numerous variables other than size and distance that affect the sales of a retail store, and that there needed to be additional techniques to identify and quantify those additional variables.

Super Valu continued with Huff's work and eventually developed the first commercial gravity model, SLASH (Store Location Analysis System Heuristically). Based on store sales and trade area performance data for hundreds of Super Valu stores, SLASH was not only used by internal Super Valu market analysts, but was also licensed to market analysis departments at other supermarket retail and wholesale firms.

The commercial gravity models in use today are based on this original work at Super Valu; all utilize comparable data and physical measurements: population, estimates of sales potential, competitive store sizes, competitive sales estimates, and the distances between and among the population centers and the retail stores serving those population centers.

Because gravity models are based on relationships involving size and distance, they are more appropriate for explaining consumer behavior patterns with respect to convenience-driven retail environments (as opposed to destination-driven comparison goods retailers). Convenience-driven retail stores are typically described as those that emphasize commodity goods—merchandise that is essentially identical from store to store in terms of quality and price.

As such, shoppers typically have no need to shop around for comparison purposes; thus, they conduct most of their shopping at only one store. Further, the store they shop tends to be relatively close to their residence. The best examples of convenience-driven stores are supermarkets. Such stores have considerable repetition in purchase patterns, and typically have trade areas that can be geographically defined in a relatively consistent fashion.

Further, because we all must eat, there is a more direct relationship between sales potential and population with supermarkets than with other retailers whose appeal may be more related to lifestyle or demographic considerations. Other retailers, especially shopping goods or specialty stores (such as furniture stores, apparel stores, and home improvement stores), tend to be more destination-oriented and draw their customers from broader geographic areas on the basis of their specialized needs and particular shopping motivations. Customers of such stores often cross-shop in order to compare quality, prices, service, and assortments; they do not shop only at the closest store. Therefore, gravity models have not been particularly appropriate for site analysis studies involving stores other than convenience-driven retailers; their applicability is most relevant for supermarkets and, to a lesser extent, drug stores and convenience food stores.

Gravity models are the technique most commonly used by supermarket chains, wholesale grocers, and consulting firms for developing supermarket sales projections. In fact, the gravity models primarily in use today were developed by supermarket, food wholesaler, or consulting firms that specialize in supermarket location research activities. These models are proprietary, and their size/distance relationships (formulas) are not in the public domain.

Gravity models today assess the geographic relationships that exist between and among stores in a trade area in terms of their sizes, their distances from one another, their distances from centers of population, their merchandising and operational practices, and their estimated sales levels. Because gravity models simulate a market or trade area, all relevant information must be accurately portrayed in the model. Such information includes the locations, sizes, and estimated sales of all stores in the area being studied, the population of the trade area broken down into geographic groupings or sectors and the study area expenditure potential, and any barriers to travel that may exist (man-made or natural). The reliability of gravity models is further dependent upon precise geographic measurements between and among all competitors, population sectors, and barriers, as well as an accurate size and sales estimate for all competitors in the trade area. In addition, solid analyst judgment is needed to assess each competitor's merchandising and operational characteristics in order to properly posture each store in the trade area being studied.

There are several steps in constructing a gravity model preparation for developing a sales forecast for a proposed store. These steps usually involve fieldwork, constructing the gravity model, balancing the model, and forecasting sales for the unit being studied. Typically, the research analyst starts with fieldwork in the trade area being studied. This fieldwork consists of several types of activities:

- Driving the area in the vicinity of the site in order to assess such site characteristics as visibility, ingress/egress, parking, retail synergy, and adjacent land use.

- Driving the major access routes in order to determine patterns of local and regional accessibility as they relate to the site.

- Defining the effective trade area to be served by the proposed store and any barriers within that trade area.

- Identifying every competitor within, and immediately beyond, the defined trade area, and then evaluating each store in terms of its size, estimated sales, merchandising and operational characteristics, etc.

- Gathering up-to-date population and demographic data from local sources, as well as any population growth statistics that are available.

Upon completion of the fieldwork, the analyst can begin to construct the model. As a first step, all relevant competitive stores serving the defined trade area are positioned at their appropriate locations on a good-quality base map. Next, the trade area on the map is divided into population sectors in order to group geographically and demographically similar population bases into smaller, manageable groups (or sectors). Ideally, these sectors contain no more than 3,500 people, all of whom are demographically homogeneous. It is important that the scale of the map be determined so that the measurements of store and population sector centroids (typically measured in inches) can be converted into miles in the gravity model.

In order to simulate the market or trade area accurately, all relevant competition and population data on the trade area map must be entered into the gravity model. This is typically done by constructing vertical and horizontal axes to the left and beneath the boundary of the trade area. The "intersection" of this point is to be used in determining the x- and y-coordinates for each point in the model: each store location, and each population sector centroid. In addition, if there are any barriers that might restrict customer mobility within the trade area (such as rivers, large cemeteries, limited access freeways, industrial belts, etc.), the x- and y-coordinates for each end of such barriers, as well as for any cross-points, are also determined.

Along with store and sector measurement information, the analyst enters population and expenditure information for each population sector in the model. Population data are generally obtained from one or more reliable sources and then assigned to population sectors. Certain demographic information is also obtained for each population sector to be used in calculating the average per-capita expenditure potential for each sector (either on a weekly, monthly, or annual basis). The multiplication of the population by that per-capita expenditure potential results in the total sales potential that exists in each sector. This is a critically important step, since the gravity model allocates sales potential among the competitors on the basis of numerous variables that are adjusted by the analyst (this process will be discussed in the next few paragraphs). Whatever sales potential is left over after this allocation is usually referred to as "float" or "leakage." A discussion on this subject follows later in this chapter.

Upon completion of the data entry (i.e., when all competition, population, barrier and expenditure data have been entered into the model), it must be "balanced" or brought into an accurate simulation of the trade area. In general terms, this process involves studying the geographic distribution of the expenditure potential in relation to the competition and barrier information, and estimating how this expenditure potential is currently being allocated to each of the competitors serving the trade area. A number of different variables may be manipulated in order to accurately simulate the allocation of the expenditure potential over the competitive environment. These variables are used to estimate the market performance of each competitor in terms of where its sales originate, the shape of its market share decay curve, the type of image it has, and so on. As any gravity model analyst will attest, balancing the model is the most crucial step in the entire sales forecast procedure, for it is in this balance that the model simulates the allocation of sales potential throughout the competitive environment.

There are several types of variables that need to be adjusted by the analyst when balancing a gravity model. The first of these is the percentage of each competitor's sales that is derived from the subject trade area. This variable is often referred to as "percent explained" or "draw." For example, a competitor who is located near the trade area boundary might have a "percent explained" that is relatively low (perhaps in the range of 15 percent to 25 percent), which suggests that only a small portion of its sales are derived from the subject trade area. Conversely, a competitor who is situated near the center of the subject trade area (and thus relatively close to the site under consideration) likely derives the greatest portion of its sales (perhaps 75 percent or more) from the defined trade area. During the field-work stage, as well as in the preparation of the data for entry into the gravity model, the analyst will make an estimate of "percent explained" for every competitor in the model.

A second variable is called "curve" or "pulling power." This variable provides an indication of the manner by which a store's sales are distributed. The higher a store's "curve," the greater its proportion of sales generated from consumers residing close to it; the lower the "curve," the more sales that are generated from consumers residing at a greater distance. In effect, the "curve" is a representation of a store's market share decay curve. Generally, the higher the "curve" value, the more rapidly a store's market share decays over distance; the lower the "curve" value, the more slowly a store's market share decays over distance. A good example would be a small mom-and-pop grocery store versus a large-format, discount-oriented supermarket. The mom-and-pop grocery store would likely have a high "curve," indicating that it derives almost all of its business from within a small distance. Conversely, the large-format, discount-oriented supermarket would likely have a relatively low "curve," suggesting that it derives a significant portion of its business from a much greater distance.

A third variable is the "density radius." Simply stated, the "density radius" is a number that indicates the geographic extent of each store's trade area and is generally a function of the population density within the area being modeled. A typical default model density radius is 2.0, suggesting that each store in the model obtains most of its sales from within a radius of 2.0 miles. In certain geographic areas in which population density is particularly low (such as one might find in a rural trade area), a "density radius" of greater than 2.0 would be appropriate; where the density is particularly high (such as in a mature urban area), a "density radius" of less than 2.0 would be more appropriate. It is up to the analyst to stipulate the density radius for every model that is constructed. In addition, there may be instances where there are significant differences in population density surrounding certain stores or in certain population sectors in a model. There may also be differences in the way different stores reach out over distance in pulling their customers. In such instances, the analyst may want to alter the density radius in certain parts of the trade area. This can be done either by varying the sector density radius (for a particular sector), or store density radius (for a particular store).

A fourth variable is called "leakage" (also referred to as "float" or "noise"). Leakage represents the sales potential dollars within the area being modeled that are not absorbed by the stores being modeled. These sales potential dollars are being garnered by numerous other retailers. Typically, recipients of "leakage" dollars include convenience stores, small mom-and-pop grocers, produce markets, meat markets, the dry grocery departments of discount department stores, and, more recently, neighborhood service stations featuring a large offering of dry groceries, dairy, and snack items. Also, there may be food potential dollars that are spent in supermarkets situated outside the trade area that are not included in the model. Because of their pulling power, these stores are able to generate some sales from the study trade area even though they are not located within, or very near, the trade area. The numerous types of stores that absorb leakage are not entered into a gravity model analysis, but they do account for a significant proportion of supermarket-type merchandise sales. For a typical trade area, "leakage" is in the range of 15 percent to 25 percent, with the modeled supermarket competitors accounting for the remainder of the available potential.

By altering one or more of the above variables in order to simulate the trade area being studied more accurately, the model is gradually brought into a balanced condition. Throughout this process, the analyst is continually reviewing the final major variable common to all gravity models—the "image" (or "flavor" or "power") of each competitor. "Image" is a number that is assigned by the gravity model to each modeled supermarket as a means of indicating its strength relative to all of the other modeled supermarkets. The sum total of all image ratings for all stores in the model divided by the total number of stores in the model yields an average image rating of 100. Therefore, a competitor with an "image" of 120 is perceived by the analyst as being better than average for the trade area being studied, while a competitor with an image of 80 is considered below average. When the competitors are appropriately ranked relative to one another (i.e., they each have an appropriate image according to the analyst), the gravity model is balanced, at which point it is ready to predict sales.

The art of balancing the model is strongly dependent upon an analyst's experience, knowledge, and good judgment, for the rank order of competitors in a trade area must be decided in terms of their image characteristics and desirability. If balancing a model were strictly mathematical, assigning an image number on the basis of each competitor's efficiency (sales per square foot performance) might seem like an appropriate way to account for the relative strengths of competitors within the trade area. However, this relationship is not as straightforward as it might seem, for there are store characteristics in addition to sales per square foot performance that suggest a store's relative strength in a trade area, including specialty departments, location type, merchandise mix, format, and so on. Therefore, while stores with higher sales per square foot generally have higher images relative to stores with lower sales per square foot, the relationship is not absolute, direct, or even proportional.

A gravity model can be used to forecast sales levels associated with any changes in a competitive environment, such as a new store opening, a store closing, a store remodel or expansion, or a store relocation. The sales projections forthcoming from the model are based on a finite sales expenditure potential. In other words, the sales projections implicitly assume that no new demand or sales potential will be created; the addition of a new store to a trade area (or remodeled or expanded existing store), results in a reapportionment of existing sales dollars. With every iteration of a balanced model in which a store change has been introduced, a sales forecast is generated for each store in the trade area—existing stores as well as any new stores or modified stores. In this manner, it is possible to determine the probable effect of one competitive change on all other stores in the trade area.

For example, imagine a trade area that is affected by eight food stores which achieve a combined weekly volume of $3.2 million. The analyst has adjusted their individual "percent explained" to the point that $2.6 million of their combined sales volume is actually derived from the defined trade area. If the estimated total weekly food potential in the trade area is $3.4 million, the eight existing food stores capture 76.5 percent of the potential ($2.6 million divided by $3.4 million); therefore, there is implied leakage of $0.8 million (or 23.5 percent). The addition of a new store to this trade area environment will result in the sales dollars being redistributed among the nine stores that will be serving the trade area; the eight original stores will give up a combined amount of trade area volume that is equal to the trade area volume projected for the ninth store.

In a few instances, such as in an underserved trade area where leakage is particularly high, the analyst may choose to "convert" some leakage. In effect, by converting leakage, the analyst believes that some of the expenditure potential that is not being captured by the major supermarkets serving the trade area (because of the limited number of shopping opportunities serving the trade area), would be retained if a new and/or superior store were to serve the trade area. In such circumstances, a portion of the sales volume to be achieved by the new store will come from retained leakage, with the remainder coming from the existing stores.

During the sales forecasting process, the analyst continually evaluates the forecasts using many of the tools provided by gravity models. Perhaps the primary means for this evaluation consists of a "market share by sector" report which shows the model's allocation of market share for every store included in the model by every population sector in the trade area. Market share is defined as a store's proportion of the available sales potential that exists in a particular trade area or sector. As an example, consider Sector A that has a population of 3,000 persons and a per-capita weekly food store expenditure potential of $40.00 (for a total weekly sales potential of $120,000). The gravity model has determined (through the balance allocation) that Food Market #1 obtains $66,000 per week (or $22.00 per capita) from Sector A. This results in a Food Market #1 market share of 55 percent in Sector A. The market share by sector report can be used to evaluate each and every food store's market share in each and every sector in the trade area. Should the distribution of projected market shares seem unreasonable, the analyst can further adjust such variables as "curve" and "percent explained," in order to bring the projected market shares to a more realistic distribution.

Gravity models offer a number of advantages to their users:

- Gravity models use a relatively limited array of data—analysts typically work only with population, demographic, and competitive information regarding a store's trade area.

- Gravity models do not require the development of a store database. All of the relationships that are needed to balance the model are either contained within the statistical equations that make up the model, or are provided by the analyst based on his fieldwork.

- Gravity model forecasts are conducted in a consistent and rigorous manner. There is a strict discipline which the field analyst must follow in order to maximize the validity and reliability of the model. After all, it is the analyst's task to simulate the trade area being studied, and such a simulation rests heavily on accurate data inputs. Conversely, while the discipline is strict, the training program for gravity model analysts typically is not very long.

- Gravity models offer an analyst the ability to conduct multiple "what-if" scenarios. Once the model has been constructed for a particular trade area, it is possible to quickly evaluate numerous alternative scenarios. For example, the sales effects of alternative store sizes can be studied, or the effect of a possible store closing can be assessed, or the effect of an expansion of a nearby competitor can be determined.

- The gravity model is an excellent tool for simulating a very large market area (e.g., a metropolitan area such as Dallas), in which the impact of numerous competitive changes can be assessed quickly. While developing such a model requires a large amount of fieldwork, and would need periodic updating, it represents a major strategic advantage to the firm that chooses to maintain such a model.

As might be expected, there are several limitations associated with gravity models. While gravity models are typically known as mathematical simulations, it is important to remember that there is ample opportunity for analyst error. Balancing a gravity model is not a mathematical phenomenon; rather, it involves quantitative input regarding some qualitative aspects of the competitive environment and the market. In other words, gravity alone (or the combination of size and distance) does not explain variations in food store sales in a trade area; the analyst must be able to adjust one or more variables to better simulate the competitive environment if they are to adequately bring the model into simulation. As a result, analyst experience and expertise play a definite role in working with gravity models.

A second limitation of gravity models concerns the methodological premise upon which the model is based. By its definition, a gravity model in large part explains store sales on the basis of size and distance—the size of a store and the distance over which it draws customers. Such a relationship, in its purest sense, implies that shoppers travel to the nearest shopping opportunity, never deviating from the motivation of locational convenience. But to the extent that other motivations enter into the shopping decision, shoppers may not go to the nearest store. In such cases, analyst judgment must enter into the process in order to properly set up the model's exceptions and better simulate reality.

A third limitation of gravity models concerns the increasing proliferation of nontraditional stores, particularly in the retail food industry. Conventional food stores were the norm in the retail food industry in the 1950s, 1960s, and even early 1970s. However, the late 1970s, 1980s, and 1990s have seen the development of numerous alternative food store formats: warehouse markets, club stores, limited assortment stores, combination food/drug stores, hypermarkets, natural food stores, upscale and specialty food stores, to name a few. Because these alternative formats usually rely on shopper motivations which go well beyond convenience of location, some of the magnetlike relationships underlying the use of gravity models are put into question. In other words, a reasonable question on the part of a site analyst might be, "How effectively can a gravity model, which is based on the overall importance of locational convenience, simulate the trade area pull of a retail food store that is driven by other, non-convenience-oriented motivations (such as quality, selection, membership considerations, etc.)?"

Finally, gravity models are limited by their need for the calculation of sales potential and for precise competitor sales data. Because supermarkets carry essentially the same type of merchandise regardless of format (dry groceries, frozen food, dairy products, meat, produce, etc.), each store in a trade area derives its business from a common definition of sales potential. But the same cannot be said for other retail formats (such as home centers, department stores, etc.), where each member of the industry carries a different, often unique product mix. As such, the calculation of a uniform sales potential for such stores is not possible. Further, because the gravity model allocates the sales of all competitors in a trade area among the various sectors that exist in the trade area, it is critically important that accurate sales estimates for each competitor be obtained. This is generally more difficult to achieve in non-food-oriented retail sectors, and thus limits the use of the gravity model to the supermarket industry.

In summary, it is evident that gravity models play an important role in location research and site analysis activities, especially in the retail food industry. With relatively few data inputs needed, coupled with relatively modest analytical training requirements, the gravity model represents a means to evaluate sites and forecast sales that is readily available to virtually all sizes of retail firms. Further, in the convenience-driven sector of the retail industry (notably its supermarket component), gravity models have proven to be accurate predictors of sales. Gravity models also allow for the contemplation of numerous alternative scenarios in a relatively short period of time once the model has been constructed and balanced. But, like any other technique for evaluating the sales potential for a proposed store, gravity models need analyst judgment if they are to adequately simulate the realities of the trade area.

---

1. Originally developed by William J. Reilly (1929), and reported in The Selection of Retail Locations, by Richard L. Nelson; published by F. W. Dodge Corp., 1958, p. 148–149.

2. Notably David L. Huff and Larry Blue, A Programmed Solution to Estimating Retail Sales Potentials, Lawrence, Kansas: Center for Regional Studies, The University of Kansas, 1964.

10 Sales Transfers

ASSUME THAT AN ANALYST FOR A HYPOTHETICAL retailer has faithfully followed all of the approaches and methodologies discussed in this book thus far. Using them, the analyst has identified a new store deployment opportunity in an existing market. An analog-based forecast for the proposed site suggests that its sales potential is quite strong. However, unbeknownst to the analyst, development of a store at the proposed site will severely cripple the sales performance of several existing sister stores. The development of a new store that "steals" most of its sales from existing stores is the real estate equivalent of "robbing Peter to pay Paul." For this reason, the location research analyst's job is not done until an assessment of the impact that a new store (or stores) will have on the existing store network has been completed. In short, the analyst must conduct a "sales transfer analysis" before a definitive recommendation regarding any site can be made.

Generally, when a retailer opens a new facility in a market where he or she has existing stores, they experience some level of sales transfer from the existing unit(s) to the new unit(s). The transfer of sales is the result of a change in customer shopping habits; a certain proportion of the established store's customers now find the new unit to be a more convenient shopping alternative. In some instances the transfer is very minimal, and can often go undetected. In other instances, the transfer of sales is so great that one, or more, existing units can be decimated by the opening of a sister store nearby. Given the potential severity of the impact that new stores can have on established stores, many retailers now realize that being able to predict accurately the impact that a new store will have on surrounding sister stores is every bit as important as being able to estimate accurately the sales performance of a new store.

The ability of a retailer to accurately estimate sales transfer often has a direct bearing on its ability to sustain healthy growth in sales and profits. This is especially true for new retail ventures. If a fledgling retail start-up sequentially opens stores that are too close to one another, the resultant severe sales transfer is not only crippling to the performance of established stores, it also sends up a "red flag" to the investment community, as the chain's comparable store sales growth will appear flat or even negative. Quick-service restaurants (and other franchise-based retailers) have also recognized the importance of accurate estimates of sales transfer. For such retailers, the ability to predict sales transfers from an existing store to a proposed unit could prevent them from becoming entangled in costly infringement lawsuits from franchisees. Finally, all retailers need to consider the potential sales transfer implications resulting from the deployment of a new store in order to determine if the proposed site, regardless of its sales potential, fits into the strategic plan, and generates sufficient "net new" sales.

The concept of sales transfer is a relatively easy one to grasp. As discussed in Chapter 3, a store's sales distribution can be quantified by conducting a survey among its customers to determine how much they spend at the store and where they live. This sales distribution can be mapped to provide a geographic depiction of a store's sales distribution. When a new sister store enters a market in an area proximate to an existing store, some of the existing store's customers invariably find the new store to be a more convenient place for them to shop. When this occurs, they transfer their patronage (and implicitly, their sales) to the new store. Convenience is usually (but not always) measured by customers in terms of the distance they must travel to shop either store. For customers residing midway between a new store and an established store (a point referred to as the perpendicular bisector), it is equally convenient for them to shop both stores (Figure 10-1). Customers who reside between the established store and the perpendicular bisector are less likely to transfer their business to the new store; the established store is more convenient for them. Conversely, customers residing between the perpendicular bisector and the new store are more likely to transfer their business, while customers that are intercepted by the new store, or would literally have to drive past the new store to shop the established store, are most likely to transfer their business to the new store.

Convenience should not be considered only in terms of distance. For example, a customer closer to a new store, may continue to shop the established store because accessibility to the established store is more efficient. It is also worth noting that some of an established store's customers will not transfer to a new store even when the new store is much more convenient to them. As an example, consider customers who have historically shopped an established store, but now have a new sister store that they must drive past if they wish to shop the old store. Most of these consumers will transfer their business from the old store to the new store. However, some will continue to shop the old store because of, for example, relationships that have developed between themselves and the staff of the established store, their familiarity with the layout of the established store, the convenience of the established store relative to other frequently visited destinations (e.g., a relative, friend, workplace, shopping), and so on. Other factors that may effect the propensity of customers to transfer to a new store include:

- **Safety.** Customers will be less likely to transfer their business to a new store if it is perceived to be located in an unsafe area.

- **Physical barriers.** Customers will be less likely to transfer their business to a new store if there are barriers which inhibit their ability to access the store (e.g., rivers, secondary roads, toll roads).

map 10—1

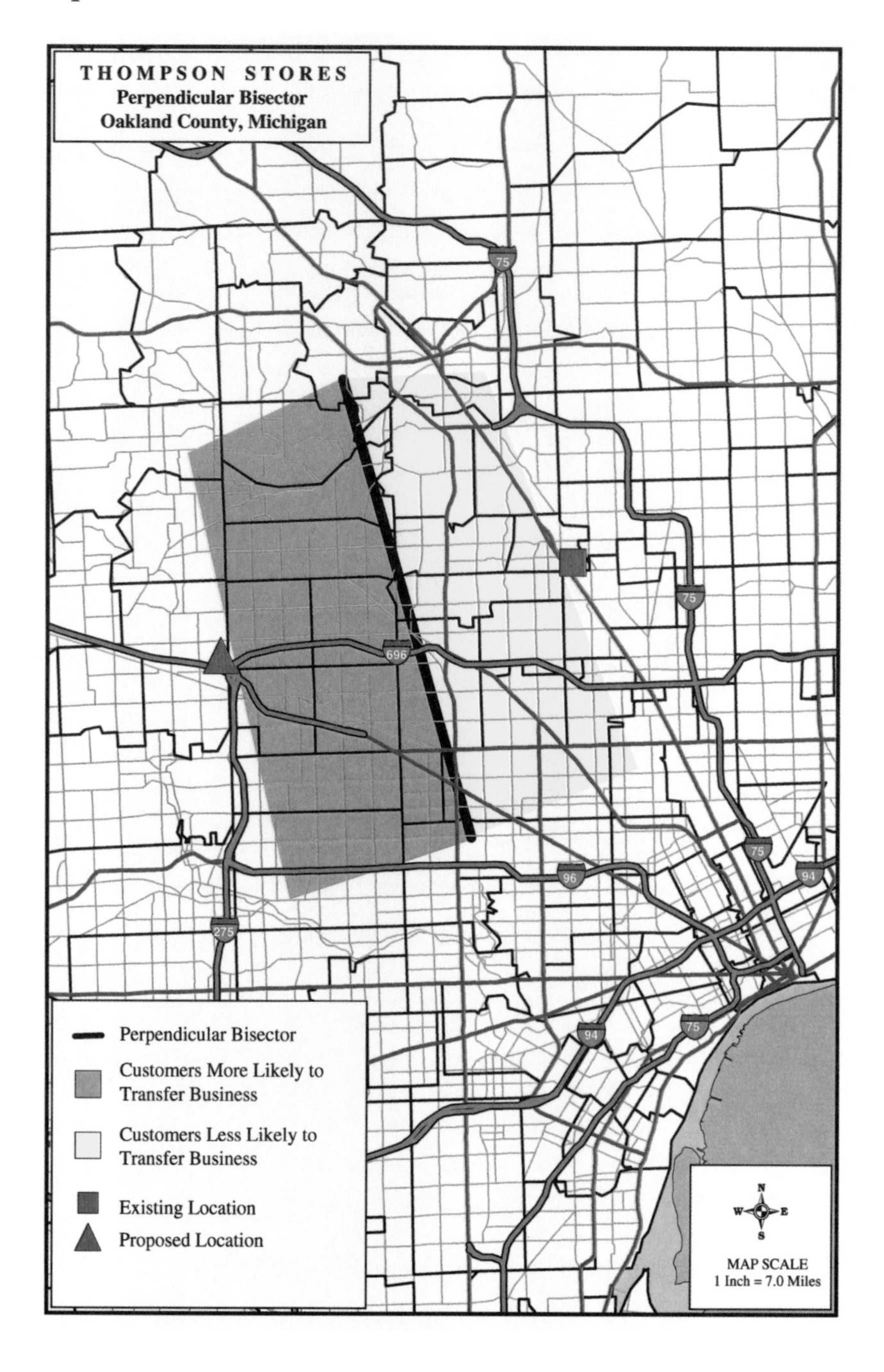
THOMPSON STORES
Perpendicular Bisector
Oakland County, Michigan
75
696
96
275
94
Perpendicular Bisector
Customers More Likely to Transfer Business
Customers Less Likely to Transfer Business
Existing Location
Proposed Location
N
W
E
S
MAP SCALE
1 Inch = 7.0 Miles

- **Sales tax.** Customers will be less likely to transfer their business to a new store if there are sales tax disincentives associated with the community or state in which the new store is located.

The preceding examples are certainly not all inclusive; anything that is an impediment to shopping will likely have an influence on the degree to which a new store will transfer sales from an existing store.

When evaluating the anticipated sales performance of a proposed new unit, it is imperative to consider it in the context of the impact it will have on existing stores. Specifically, the sales forecast for the proposed store is added to the anticipated sales of existing stores net of sales transfer. This total is compared with the sales performance of existing stores prior to the deployment of the proposed store. The incremental sales are referred to as net new sales, and provide an indication of the magnitude of the net sales gain that will result from the deployment of the proposed new store. The level of net new sales being generated is the truest measure of the success of the new location.

Before discussing the analytical process used in conducting sales transfer analysis, we want to interject a word of caution to the analyst regarding possible sales transfer pitfalls. As the vernacular implies, a sales transfer analysis seeks to estimate the impact a new store will have on an existing store's sales. As such, the analyst is advised to avoid the temptation to base this analysis on the distribution of customers, as the two distributions (i.e., customer versus sales distribution) can differ dramatically.

Another potential sales transfer pitfall is the failure of the analyst to incorporate differences in sales distributions attributable to multiple customer components into the transfer process. For example, retailers such as computer superstores, or membership warehouses, serve not only traditional consumers, but also businesses; usually, the sales distributions for these customer bases differ significantly. As such, a new sister store may impact one of the customer components severely while having only a negligible impact on the remaining customer component.

In order to quantify the sales transfer differences that will result from the deployment of a new store, the analyst must be capable of retrieving and mapping separate sales distribution data for each customer component. The analyst can then easily determine the extent to which an existing store's sales for each customer component will be impacted by a new store entering the market.

A retail segment for which multiple customer components is particularly problematic is the quick-service restaurant industry. Generally, quick-service restaurants do not have a convenient means to collect information for their multiple customer components (e.g., consumers visiting from home, from work, while shopping, etc.). Several factors make the collection of this data particularly difficult for this industry:

- The trade areas of quick-service restaurants are usually quite small. Therefore, unlike many other retailers, quick-service restaurants cannot use a point of sale (POS) system to collect customer ZIP Codes for the purpose of studying the distribution of its sales; ZIP Codes are simply too large a unit of geography to be used for this analysis. In fact, it is not uncommon for a quick-service restaurant's entire trade area to be located completely within one or two ZIP Codes.

- A system used to collect sales distribution information for quick-service restaurants must be capable of distinguishing between the sales distribution of several different customer components. Usually, point of sale (POS) systems are not flexible enough to accommodate this need.

Because of these problems, the collection of accurate sales data for quick-service restaurants is typically accomplished via customer exit surveys at the point of sale. The most important step in this process is the development of a questionnaire addressing the trip pattern which resulted in a purchase at the restaurant. Customers are typically questioned about their trip origin, trip destination, and which of these two points was the primary generator of their trip to the restaurant.

For example, a customer may visit the restaurant from home and return home after eating; therefore, this sale would be designated a residential sale and assigned to the address of the customer's home. What if the customer's visit originated from home but the ultimate destination is work? The questionnaire must be designed in such a way as to establish which of these two points was the generator of the customer's visit. If the restaurant is only blocks from the customer's office, but miles from their home, then the customer would, most likely, advise the interviewer that the trip was generated because the restaurant was most convenient to the place of work, and the sale would be attributable to the work place and not the customer's place of residence. As this example illustrates, the "trip pattern" issue can be very complicated if the survey questionnaire and process is not well conceived.

Finally, a fairly common pitfall when conducting a sales transfer analysis is in not recognizing the strength in drawing power that different store prototypes have, and in failing to incorporate these differences into the sales transfer analysis. For example, a new 200,000 square foot Wal-Mart supercenter will have a significantly greater transfer impact on an existing 85,000-to-100,000-square-foot prototype than it would on a contemporary supercenter sister store. Similarly, the sales transfer impact on a conventional freestanding quick-service restaurant will be different depending on whether the impacting unit is another conventional sister store, or an alternative prototype such as a double drive-thru or a truck stop location.

Sometimes all an analyst can do to address this issue is use common sense (e.g., the existing store is old and small; therefore, the sales transfer will likely be more severe than usual). Ideally, the analyst can draw upon actual instances where two differing prototypes have competed against each other as the basis for projecting what will happen in future comparable situations.

The preceding text has focused on some of the shortcomings typically encountered in estimating sales transfers. The balance of this chapter will discuss how to estimate the proportion of sales that will likely be lost when an existing store encounters competition from a new sister store.

Many of the methodologies presented in this book draw upon the theory that the best means to predict the future is to study and quantify what has happened in the past. The same holds true for sales transfer analysis; the best means of estimating what future sales transfers will be is to examine and quantify what they have been in the past.

Every retail chain that is actively opening new stores in markets where it has existing facilities should create a "before" and "after" database of stores that were impacted by new units. A historical analysis of each impacted store's sales distributions by geographic segment (e.g., ZIP Code, Census Tract, grid) pre- and postsister store openings allows the retailer to track the percent of sales lost to the impacting sister store network not only overall, but on a segment by segment basis. With enough data from an array of stores at varying distances from one another, it is possible to generalize about the proportion of sales that will be lost for each segment, considering its position vis-à-vis the perpendicular bisector, the existing store, the new store, and so on. While this is the simplest and most direct way to calibrate sales transfer impacts, it is only one of many possibilities. In general, if spatially disaggregate sales data is available for a sufficient number of pair-wise interacting stores, transfer functions can be readily calculated.

As with the sales forecast process, the calculation of sales transfer is most accurate when undertaken on a disaggregate basis. A careful analysis of the existing unit's sales distribution by segment (e.g., ZIP Code, Census Tract, grid, etc.) should be undertaken. This allows the analyst to identify all segments that will be impacted by the proposed unit, and assess the degree to which each segment will be impacted based on its situation relative to each store. As mentioned earlier, each segment will be impacted differently by the introduction of a new unit in the marketplace, depending on its distance and accessibility to the new store.

The first step in estimating the sales that will transfer from an existing store to a proposed sister store, is to map the actual sales distribution for the existing store by segment. Once the sales distributions for an existing store, or stores, are retrieved and mapped, the process of conducting the actual sales transfer estimates can commence.

Assuming that the area located between the new and existing stores is relatively uniform, has roughly equal accessibility throughout, is void of any physical and cultural barriers, and so forth, it is plausible to expect that the greatest percentage of sales transfer would occur in areas that are farthest from the existing store and closest to the proposed store. Conversely, the smallest percentage of sales transfer would occur in the areas immediately around the existing store and in those areas where consumers would have to drive past the existing store to shop the proposed store. While these assumptions are intuitive, the challenge for retailers has been in identifying how severe the strongest percentage impact should be, how modest the weakest impact should be, and how this rate should vary as the distances from the existing and proposed stores vary.

The rate at which transfer occurs varies depending on the type of retailer. Further, even for the same retailer, factors such as those noted above (e.g., customer shopping habits, physical and cultural barriers, accessibility, safety) render simply applying a consistent pattern of transfer percentages to segments over geography inaccurate; transfer percentages must be adjusted upward if a proposed site has superior access, downward if a proposed site is considered unsafe, and so on. In short, there is no simple formula which can be applied across all retailers, or even among the same retailer, for all situations; therefore, each retailer should study the actual transfer impacts that have occurred within each segment for existing stores that have been impacted by new stores in order to determine the transfer rates unique to their operation. Armed with this information, the retailer will be capable of generating accurate estimates of sales transfer.

A word of caution regarding several of the "canned" transfer methodologies that are commercially available to retailers today. In addition to the concerns raised in the preceding text (e.g., no approach should be universally applied, unique situations such as access, safety, and barriers need to be incorporated into the analysis), several of these approaches are fundamentally flawed because they only consider the sales transfer that will occur in the area where each store's trade area overlaps. The inherent weakness of this approach is that it ignores any sales transfer that may occur outside the trade area overlap area. Map 10-2 illustrates the difference between the trade area overlap approach versus an analysis of sales transfer by segment, as endorsed herein.

map 10—2

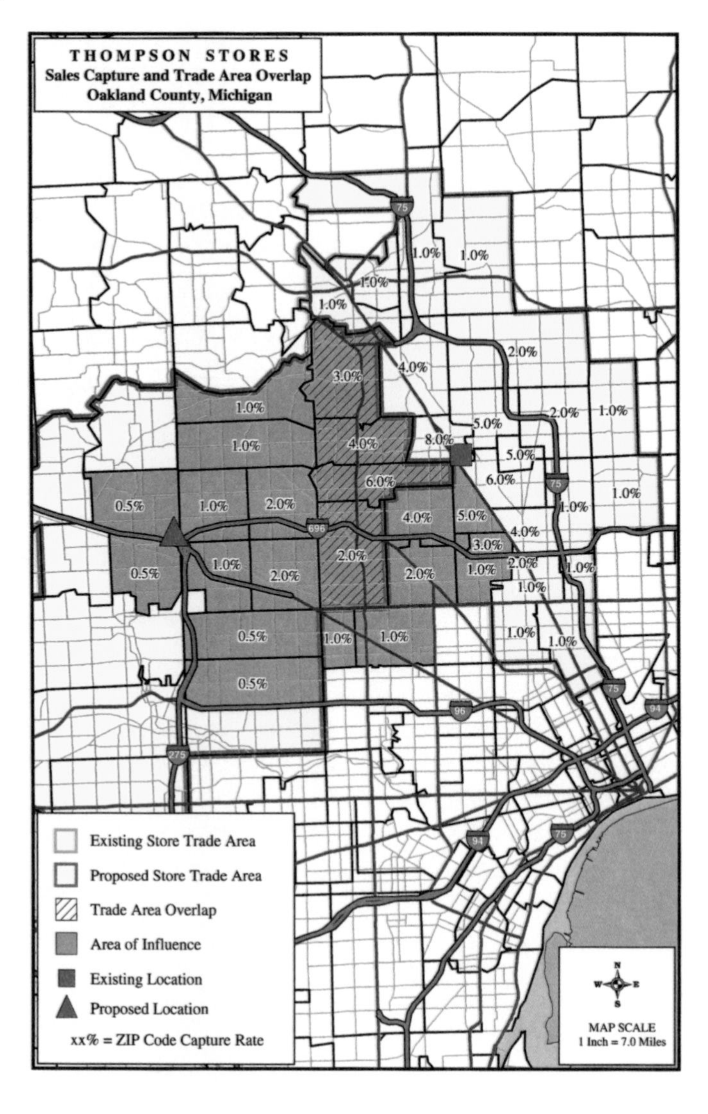

The example presents the percentage of total sales that an existing store is achieving from each segment (i.e., ZIP Code) surrounding the store. The existing store captures sales from segments adjacent to and beyond the location of the proposed store. A transfer analysis including only the trade area overlap for the two units suggests that the proposed unit will have an influence on only 15 percent of the existing stores sales, while a transfer analysis including all market segments indicate that the proposed unit will influence 42 percent of the existing stores sales. Thus, focusing exclusively on trade area overlap when estimating sales transfer, represents a fundamentally flawed methodology that is unlikely to yield accurate transfer estimates.

The preceding discussion pertains to all retailers. However, retailers who rely on any of the commercially available gravity models to assess sites have an alternative to the approaches described thus far. More specifically, gravity models intrinsically provide an estimate of the sales transfer from existing stores to proposed sites, as a component of the process of projecting sales for the proposed site. This is because gravity model forecasts are predicated on a reliable estimate of total sales potential for each segment within a study area. In most circumstances, this sales potential is largely captured by the existing stores. When a new site is introduced into the study area, its sales largely come at the expense of the existing stores. Therefore, with gravity models, sales transfers are estimated not only for sister stores, but for every competitor that will be impacted by a proposed store.

As this chapter suggests, sales transfer analysis should be a component of the research that is routinely conducted by a retailer when assessing the viability of a new site in an existing market. Transfer analysis not only provides the retailer with information that is fundamental to the site go/no go decision-making process (such as an estimate of net new sales), it also provides practical information which will help the impacted store prepare for its sales loss (such as what are the staffing, inventory, service, etc. implications that will result from a loss of sales to a new sister store). A retailer that strives to accurately estimate sales transfer, as well as the sales performance of a new unit, will make fewer brick and mortar mistakes and will have to deal with fewer unpleasant surprises.

11 Cautions in Putting Sales Forecasting Tools to Work

THE FORECASTING TECHNIQUES DESCRIBED IN CHAPTERS 8 and 9 provide an overview of the basic sales forecasting methodologies available to retailers today. While the various forecasting techniques have all been proven effective, there always exists the opportunity to inadvertently misuse (or abuse) a forecasting technique. The purpose of this chapter is to review some of the more common concerns that arise when generating sales forecasts.

There are several issues that will be covered in this chapter. Forecast convergence, whereby two or more forecasting techniques are used to arrive at the same projection independently, and enhance the analyst's confidence in the resulting projection are discussed. Next, the chapter addresses some of the more common concerns that arise when using analogs, normal curves, and gravity models, followed by an overall assessment of selected statistical techniques. The chapter also addresses how factors such as maturity (or sales ramp-up over time) and seasonal residents should be handled when conducting sales forecasts. Finally, the appropriate use of a sales forecasting system and the issue of forecast accuracy are discussed.

Forecast Convergence

Forecast convergence refers to the use of two or more independent methodologies to generate a sales forecast for a proposed retail location. As an example, an analyst could generate a sales forecast for a proposed retail location using analogs, and then replicate the same sales forecast using normal curves. Forecast convergence is perhaps the single most powerful technique for increasing an analyst's confidence in the accuracy of a sales forecast. The logic behind convergence is similar to that of an intelligence network: if two or more agents independently uncover the same information from different sources, there is a much greater confidence that the material obtained is accurate. As with each of the basic sales forecasting methodologies that may be employed, verifying a sales forecast with convergence can be undertaken at either the aggregate or disaggregate level. That is, an analyst can use an independent method to develop a sales forecast for an entire trade area (aggregate forecasting), or for each component of a store's trade area (disaggregate forecasting). While an aggregate-level review may serve as a quick overall check to confirm whether or not the sales forecast is in the "ballpark," disaggregate analysis is typically the preferred technique. Disaggregate forecast convergence (or developing independent sales forecasts for each individual trade area sector) provides the dual benefit of both confirming an initial forecast and also exposing any portion of the sales forecast that represents a source of concern.

Table 11-1 and Map 11-1 provide an example of forecast convergence using a regression model and normal curves. In this example, the analyst has generated a sales forecast for a Kinnick's Sporting Goods store in the Tampa, Florida market. The trade area defined by the analyst incorporates 14 ZIP Codes, each of which has been forecasted independently using both techniques.

Overall, there is a significant (14 percent) difference in the trade area sales forecast generated from the two techniques prior to considering sales originating from consumers residing beyond the trade area. Because the comparison was conducted on a disaggregate level, however, it is apparent that most of the difference between the two sales forecasts is attributable to two ZIP Codes:

Table 11-1
Kinnick's Sporting Goods
Sales Forecast Comparison (Regression vs. Normal Curve Forecast)

					Regression-Based Forecast		Normal Curve Forecast	
Trade Area ZIP Code	Driving Distance	1999 Projected Population	1996 Median Household Income	1996 Median Age	Projected Per-Capita Sales	Projected Total Sales	Projected Per-Capita Sales	Projected Total Sales
33614	1.3	22,309	$29,982	35.3	$70	$1,561,630	$72	$1,606,248
33607	3.1	16,297	$19,623	35.5	$51	$831,147	$50	$814,850
33634	3.1	22,109	$40,774	35.5	$48	$1,061,232	$50	$1,105,450
33609	3.6	29,606	$34,682	42.1	$45	$1,332,270	$48	$1,421,088
33603	3.9	14,235	$25,782	35.8	$47	$669,045	$46	$654,810
33604	4.9	9,017	$24,044	34.7	$40	$360,680	$38	$342,646
33629	5.2	19,250	$50,751	41.7	$33	$635,250	$35	$673,750
33602	5.7	34,684	$12,507	33.7	$34	$1,179,256	$32	$1,109,888
33618	5.7	44,594	$49,045	36.5	$26	$1,159,444	$32	$1,427,008
33615	6.0	19,266	$39,541	36.6	$32	$616,512	$29	$558,714
33606	6.5	38,552	$35,829	37.1	$24	$925,248	$27	$1,040,904
33624	6.8	41,194	$52,894	35.5	$8	$329,552	$26	$1,071,044
33612	7.0	22,695	$24,393	34.1	$23	$521,985	$25	$567,375
33625	7.5	42,571	$47,867	33.8	$6	$255,426	$22	$936,562
Trade Area Total		376,379	$37,228	36.2	$30.39	$11,438,677	$35.42	$13,330,337
Projected Sales from Beyond the Trade Area						$3,000,000		$3,000,000
GRAND TOTAL						$14,438,677		$16,330,337

33624 and 33625. Forecast convergence therefore not only identified an overall concern with the initial regression-based sales forecast, but pointed out the individual ZIP Codes that represent the most significant cause for concern. In this example, a more detailed evaluation of the ZIP Codes in question reveals that the normal curves are understating the effect of competition—in this market, Bressor Sports (the only significant competitor affecting the site) is a particularly strong operator and has a pronounced impact on the two ZIP Codes in question (refer to Map 11-1). Because the regression-based model accounts for this particular competitive strength (and the normal curves do not), the analyst determines that the regression forecast represents the more accurate sales forecast.

If a sales projection generated with one approach converges with another, the initial reaction is positive. The findings are typically accepted, and the sales forecast and related location decisions are considered reliable. Nevertheless, it is still appropriate to subject the sales forecast and transfer impacts to a final "common sense" test. Does this sales forecast make sense? If the sales forecast produces the highest (or lowest) sales volume for this market or region, is that

map 11—1

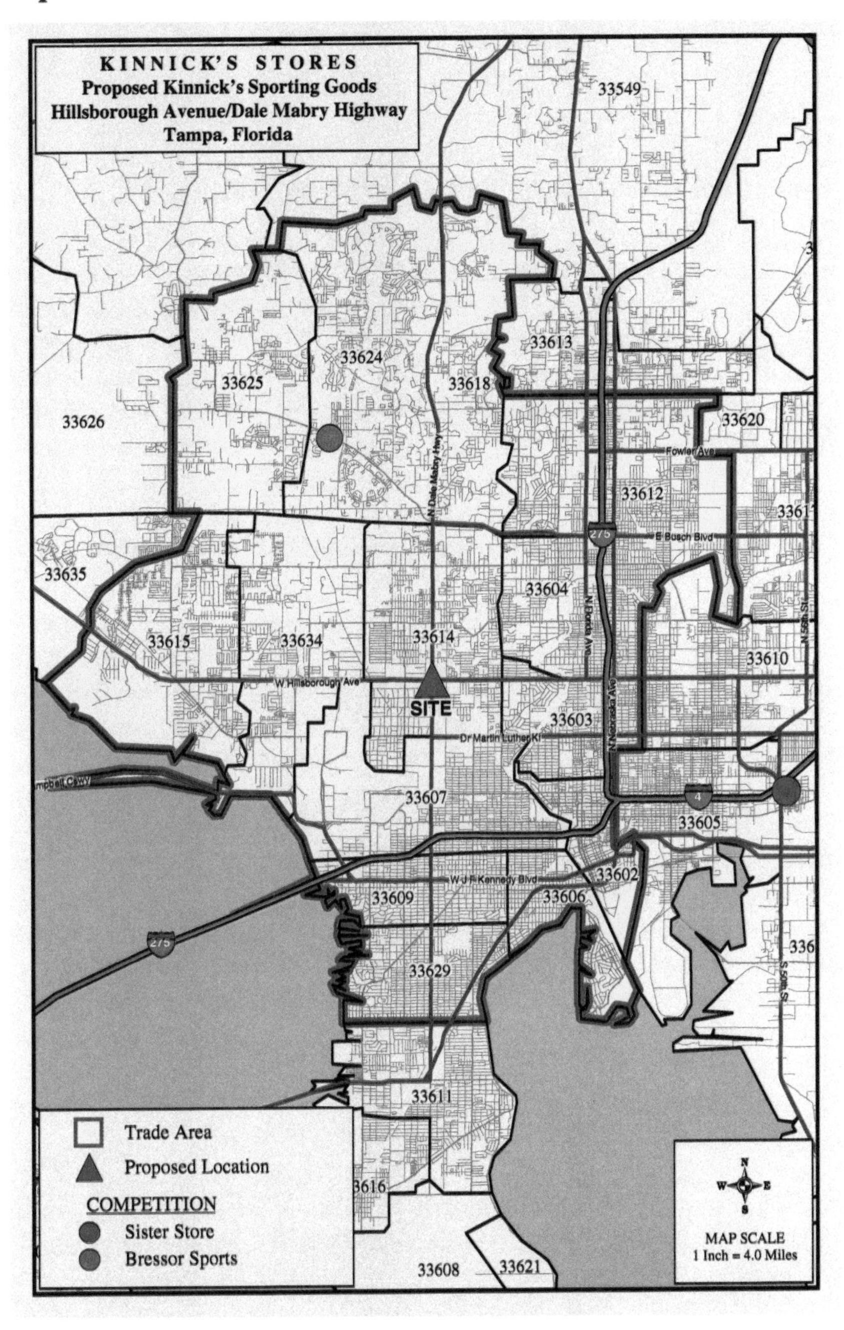
KINNICK'S STORES
Proposed Kinnick's Sporting Goods
Hillsborough Avenue/Dale Mabry Highway
Tampa, Florida
33549
33613
33624
33625
33618
33626
33620
Fowler Ave
33612
N Dale Mabry Hwy
275
E Busch Blvd
33635
33604
N Florida Ave
33615
33634
33614
33610
W Hillsborough Ave
SITE
33603
Dr Martin Luther Ki
N Nebraska Ave
33607
33605
33602
W J F Kennedy Blvd
33609
33606
275
33629
33611
Trade Area
Proposed Location
COMPETITION
Sister Store
Bressor Sports
N
W
E
S
MAP SCALE
1 Inch = 4.0 Miles
33608
33621

appropriate? Are there any regional, competitive, or other considerations that would cause the research analyst to question or doubt the forecast? If the analyst is still comfortable with the sales forecast after answering all these questions, then he or she can proceed accordingly.

Cautions Regarding Normal Curves/Analogs

One major problem frequently encountered in the use of either analogs or particularly normal curves as a forecasting technique is that they can inadvertently direct a projection toward a norm or "average forecast" rather than an extreme forecast (either high or low). There can be a myriad reasons for this, including the length of time in a market; relative competitive strengths and weaknesses; cultural biases, an analyst's reluctance to generate an unduly high or low forecast; and other considerations. Despite this tendency, if an extreme forecast is warranted and if a capable analyst can justify it, then the extreme figure should be used. If normal curves, in particular, press the analyst toward an average rather than an extreme sales estimate indicated by another technique, the analyst should remember that normal curves inherently represent an "average" level of performance. Avoiding forecasted extremes when they are warranted (a mistake common among inexperienced analysts) can result in a weaker than predicted actual performance for poor sites (i.e., the store will generate even lower sales than the already low sales forecast), or stronger than predicted actual performance for favorable sites (i.e., the store will generate even stronger sales than the favorable forecast). Obviously, while it is easier to deal with situations where actual sales exceed projections, neither scenario is desirable. Conservative forecasts resulting from a reliance on averages could cause retailers to pass on selected opportunities because they have underestimated the sales potential associated with the site. Conversely, overprojections resulting from an overemphasis on averages will yield stores with disappointing sales and profit performance.

An obvious follow-up question is: "How does the analyst know when to disregard the red flags raised by normal curves during the convergence process, and to stick with a more extreme forecast (very high or low) suggested by the analogs or regression model? Unfortunately, there is no simple answer to this question—if there were, it could simply be incorporated into a sales forecasting methodology to begin with. Some of the factors that can help to ensure that a forecast that appears "extreme" is appropriate are presented below:

- Ensuring that appropriate analog stores are used (selecting those analog stores whose demographic, competitive, and site-related characteristics most closely resemble the proposed site under consideration); if the analog stores closely match the site being forecasted, it is likely that forecast extremes suggested by the analogs are warranted.

- The presence of trade-area demographic characteristics that are unusually strong (or weak) for the concept in question suggest that forecast extremes are likely warranted.

- Providing due consideration for competition (being careful not to overemphasize weak competition while concurrently giving adequate weight to strong competition) will ensure that forecasts do not over- or understate the presence or absence of trade area competition.

A similar dilemma is faced when a retailer plans to enter a new market or new region. Retail performance in different areas of the country can either be very similar or very dissimilar; variations by region often seem almost impossible to predict in advance. Given comparable situational, demographic, and competitive characteristics, will proposed stores in Dallas perform as well as the existing Seattle division? In some instances, variations in performance may not be quantifiably explainable, requiring the analyst to use critical judgment in his or her approach to the data and its inferences.

A common mistake associated with the analog method is that the analyst may look for many matches, often at the expense of the comparability of the analog matches selected. For example, an analyst might identify as many as 50 to 100 analog matches for a given trade area or individual sector. This list should be carefully pruned to include only those analogs (perhaps 5 to 10 in total) that most clearly replicate the situation being forecasted. Tightening the tolerance of the analog variables (e.g., cutting the median household income levels from ±20 percent of the sector under consideration to ±10 percent; or excluding those analog stores that are situationally dissimilar) can assist in sharpening the sales forecast by focusing only on the best matches.

Cautions Regarding Gravity Models

Most retail gravity models utilize a variable sometimes known as "power," "flavor," or "image"—that is, the variable that represents differences in sales performance levels once distance, drawing power, and competition are taken into account. A research analyst should avoid inherently assuming that a proposed store will be "one of our very strongest," without some corroborating evidence for this assumption. While such assumptions may well be true, management claims should be supported with clear evidence if the related quantification and judgments are to enter into the sales forecast. In short, the analyst should not display any inherent bias toward any final forecast estimate before it is generated, lest the resultant sales forecast simply represent management expectations rather than an objective assessment of sales potential.

Gravity model forecasts are also critically dependent (even more so than regression or analog forecasts) on using accurate base information. Gravity model sales forecasts are inherently a "zero-sum game"—the sales forecast for a new retailer is largely dependent upon carving out market share from existing operators. As a result, if the underlying demographic and competitive information is not accurate, then the resultant sales forecast will be inherently flawed as well.

Statistical Analysis

When confronted with the results of a regression analysis or other statistical technique, many retailers often respond with one of two opposite, and equally dangerous observations.

> "This conclusion is based on the most sophisticated, mathematically rigid technique available, and therefore, must be correct," or:
>
> "This is nothing but statistical mumbo-jumbo that isn't worth the paper it's printed on."

The first group of retailers should recognize that detailed mathematical processing of variables, while highly useful, does not always represent the optimal approach to sales forecasting problems. The same cautions that apply to other techniques—namely, a need for a reality check—apply equally to regression and other statistical analyses. It is all too easy for the results of a regression analysis inadvertently to mislead a forecast analyst, particularly when the retailer in question is encountering unique markets or situations which are not replicated in the existing database. This can be attributable to the inadvertent misuse of a regression package or other statistical technique (i.e., attributing variations in sales levels to one variable when another variable is actually responsible), or by applying the results in a manner that does not make pragmatic sense, as demonstrated in the following example.

A database for the retailer Stay-Low Stores demonstrates a strong positive correlation between sales performance and median household income levels (once the impact of distance is factored out), as depicted in Figure 11-1. However, of the median household income observations lie within a range from $32,000 to $88,000. Even though there are no observations for areas with median household income levels below $32,000, the mathematically derived regression line suggests that the average annual sales per household will be $12 in areas with median household income levels of $20,000. Conversely, the regression analysis infers that areas with a very high median household income level (i.e., $120,000) will generate average annual expenditures of $90 per household. While this conclusion is mathematically logical, it would be clearly inappropriate for an analyst to accept it at face value. In this instance, the mathematical conclusion should be contrasted against a pragmatic evaluation or corroborated with supporting evidence. This is not to suggest that the analyst should reject the possibility that areas with median household income levels of $20,000 will generate sales of $12 per household; instead, there is an inherent weakness in arriving at that conclusion, and recognize that additional research is necessary to support or reject the conclusion. An unchecked or "blind faith" use of regression can result in misleading conclusions.

figure 11—1

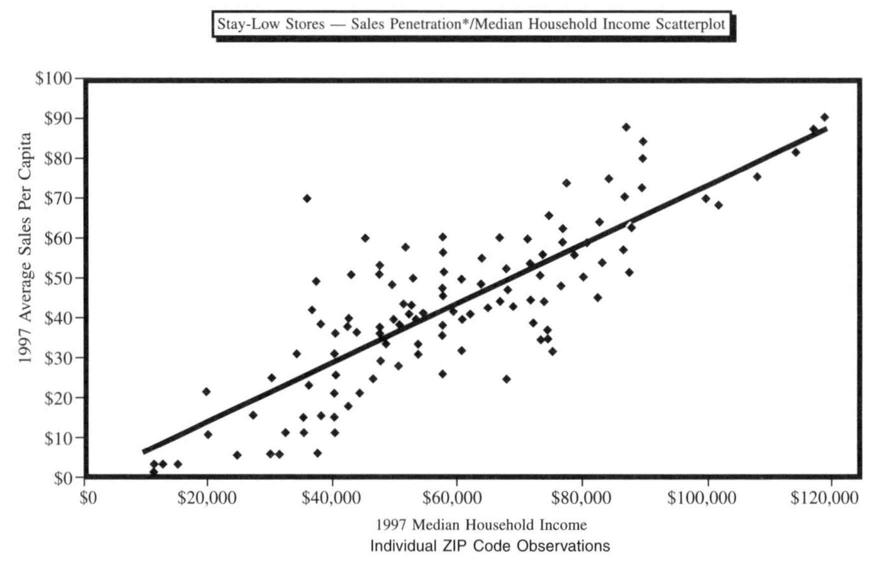

*Sales Penetration levels have been normalized to reflect consistent distance measures.

Research analysts should avoid an implicit acceptance of "black box" reliability—the inference that the results of a forecast are correct if the methodology is mathematically or otherwise accurate. Formula forecasting (a simplistic reliance on a formula for estimating sales) can be devastatingly misleading. As a case in point, mathematically correct values with a regression model can suggest that customers will travel negative distances to sites or generate "negative sales" for a proposed store—mathematical accuracies, but irrational phenomena for retail location analysis.

The response of the second group (that regression analysis is nothing more than statistical mumbo-jumbo) is more characteristic of those operators that have come up through the retail "School of Hard Knocks." As with the first group, there is a strong caution for the latter. While erroneous inferences can result from a technically correct but pragmatically naive mathematical or statistical processing of data, an out-of-hand rejection of statistical conclusions fails to recognize the powerful contribution that statistical analyses can make. For this group of retailers, the admonition is that statistical techniques can be extremely useful and reliable, and to categorically ignore the resultant conclusion may inadvertently result in less than optimal location decisions.

Often, retailers in this second category will reject the results of a statistical analysis because, in some way, they conflict with their experiences or preconceptions. This is analogous to someone rejecting the notion that smoking causes lung cancer, simply because he or she had an uncle who smoked two packs a day until he was 87. Rather than directly confronting the retailer's opinions, which often have considerable merit, an analyst is better advised to evaluate the underlying data in the context of the retailer's observations. As an example, the operator of a chain of video stores might be convinced that locating next to a supermarket chain results in improved sales performance. Rather than talking about "R-squared values" and "confidence levels," the analyst should first evaluate the underlying data to understand the basis for the operator's belief. It may be that, coincidentally, most of the video stores located next to supermarkets are also located on high traffic arteries with an absence of nearby competition. By explaining this situation and demonstrating that those video stores located next to supermarkets that do encounter nearby competition perform no better than similarly competitive stores without an adjacent supermarket, an analyst is much more likely to gain the retailer's trust.

It is also important for the analyst to remember that a skeptical operator's preconceptions and opinions are, in all likelihood, more likely to be right than wrong. Most retail executives have gotten as far as they have because they do make the right decisions more often than not. A statistical result that flies in the face of conventional wisdom should not be categorically rejected, but should be subjected to careful scrutiny to better understand whether there are some other underlying factors at work before being accepted at face value.

Maturity

Maturity (or "ramp-up") for a store occurs when it reaches its anticipated long-term sales volume, given relatively stable competition and trade area demographic characteristics. While the absolute maturity rate can be different for the same type of store within a retail chain, maturity patterns are generally characterized by a similar time interval that reflects: (1) the type of retail unit; (2) the frequency of shopping; and (3) whether there is preexisting familiarity with this type of facility and/or this particular operator in a given market.

When an existing supermarket chain opens a new store in an established market, it can generally expect to have a very rapid maturity, perhaps only taking a few months to achieve mature sales levels. Other destination-oriented stores, such as membership warehouse stores, can have a maturity period that easily extends several years. For the latter store types, especially in a new market, there are many impediments to rapid maturity, or acceptance by area consumers. Membership warehouse stores typically do little advertising, often take secondary retail locations with poor visibility, are usually not shopped on a frequent basis, and impose a membership fee (which is not an incentive for initially attracting customers). These facilities can achieve less than one-half of their mature sales potential during the first full year of operation.

Another convenience-based retail category that might be expected to mature quickly, but typically does not, are drugstores. Consumers tend to patronize the same drugstore when refilling prescriptions, so that changing to a new store to fill prescriptions involves an added effort. Although other components of a drugstore, such as the general merchandise or health and beauty aids departments, may have quicker maturity periods, the prescription counter acts as an impediment to overall rapid maturity.

Given the variation of maturity rates for different types of facilities, it is critical for an analyst to determine whether the goal of a sales forecasting system is to project first year sales, mature sales, or both (as is the case for many retailers). Inaccurately estimating maturity levels can contribute as much to the inaccuracy of a first-year sales projection as under defining a trade area or failing to account for a competitor; as such, an analyst should carefully approach the whole issue of maturity.

Seasonal Residences

Accurately estimating the impact that seasonal residences will have on retail sales performance can be a challenging task, but is critical in resort areas. Often, the incremental sales volume generated during the peak season can "make or break" the sales forecast from a pro forma perspective, making it important to accurately estimate seasonal demand. An analyst's conclusions concerning the likely impact of seasonal population trends on sales levels can also have significant consequences relating to merchandising, inventory levels and staffing needs.

There are two principal issues involved with quantifying the impact of seasonal residents on sales performance levels:

- How many seasonal residents live in the trade area during the peak season, and how long does the peak season last?

- How likely are the seasonal residents to patronize the store in question?

The techniques for estimating the number of seasonal residents and the length of the peak season are covered in Chapter 2. While there are a variety of sources that can be used to estimate seasonal population influxes, it generally represents a more challenging endeavor than estimating year-round population, and requires careful research.

The second consideration—the inclination of seasonal residents to patronize the operator in question—is of equal or even greater importance. Obviously, consumer shopping and patronage habits very widely between the bulk of the year (when living at home) and those few weeks or months when one is on vacation. Expenditure patterns for selected retailers, such as restaurants and gift shops, can increase significantly during peak seasonal periods, while expenditures on other items may be considerably less.

The best method to gain an initial understanding of seasonal resident expenditure patterns is to evaluate other stores in seasonal areas, and determine the extent to which their sales increase or stay steady during peak seasonal periods. If sales information for individual retail stores in seasonal areas is unavailable, an analyst can utilize aggregate monthly sales information, either from the U.S. Department of Commerce or from state commerce or sales tax departments, to estimate the extent to which overall expenditure patterns for the merchandise in question fluctuate over the course of the year.

Expectations Regarding Forecast Accuracy

When evaluating the various site location research methodologies, retailers invariably ask the question "How accurate are these approaches?" This would seem to be a reasonable question; whether developed in-house or by a consultant, the most sophisticated site evaluation systems can require a significant commitment of resources to develop and apply. When assessing the accuracy of any site location system, this assessment should be based more on whether or not the decision to deploy new stores was a good decision, and less on a comparison of actual versus projected sales for a location.

Many retailers forget that the goal of location research is to open successful stores not to generate accurate forecasts. Such retailers equate the site location research process with sales forecasting. In fact, it is but one of a continuum of tools that are used to help retailers open successful stores. After all, the sales forecast is merely a means of measuring the viability of one site relative to others.

A pragmatic reason for basing forecast accuracy on the success of new store openings (rather than a comparison of actual versus projected sales) is that expecting every new store to perform within, for example, 10 percent of projected sales is unrealistic. This is because no forecasting methodology can account for every factor that routinely impacts store performance.

One of the biggest impediments to forecast accuracy is a small database of stores (e.g., 20 or less stores). To understand why this is so, we offer the following two scenarios. Consider a small but successful ten-store chain. The company founder managed the first store, and has had a strong hand in the management and operation of the subsequent nine units, showering them abundantly with tender loving care. It is likely that forecasts derived from a database comprised of these stores would (at least initially) overstate the sales potential of future stores as, with every additional store, the ability of the founder to have day to day input becomes diluted. Thus, Stores 11 through 20 will not be as well managed as Stores 1 through 10, a factor that can only subjectively be accounted for during the forecast process. Ultimately, as the chains management team gains experience, store performance will likely improve and fulfill the anticipated sales performance. In this instance, it is unreasonable to conclude that the site selection system did not work. On the contrary, the new stores have been well positioned based on the application of the system, and it is now the retailer's responsibility to get management "up to speed" in the new stores so that the trade area sales potential can ultimately be realized.

Another example of forecast volatility resulting from a small database can occur when a chain expands from for example, northern California to the Midwest. Such a chain will encounter new competitors, differences in consumer tastes and preferences, different market sizes and accessibility within markets, and so on. As the forecasts will be based on the competitive environment, consumer tastes, access, and so on, in northern California, the likelihood that the Midwestern stores perform at projected sales levels is reduced. However, because the sites will have been selected based on the customer and competitive profiles determined from the database, the new stores will likely be well positioned to serve the market, and the probability that the Midwest deployment will be successful is significantly enhanced.

As the size of a site selection database increases, the volatility of forecasts generally diminishes. This is because with larger databases, the likelihood that a forecast will be generated for a situation never before encountered is greatly reduced. However, even with the largest of databases, there are factors that can cause discrepancies between actual and projected sales which cannot be accounted for in the forecasting process. Something as simple as unusual weather patterns can undermine a forecast. This is a double-edged sword. For example, heavy snowfall may result in depressed sales for the database stores. These depressed sales are implicit in any forecast generated using this database. Conversely, heavy snowfall may depress the performance of new stores whose forecasts were based on a database which did not reflect bad weather.

Other factors which are typically not reflected in a forecast but can significantly influence actual sales performance include:

- **Marketing**. Providing a new store with unusual levels of marketing support (either high or low) will significantly influence store performance.

- **Management**. Retailers generally believe that the strength of store management can influence store performance by 15% to 30%. Since it is unlikely that the store manager will have been selected at the time a forecast is conducted, it is inherently not factored into projections. A strong store manager will usually generate sales which beat a forecast and vice versa.

- **Merchandise changes**. Additions or deletions of merchandise in new stores that are not inherent in the data-base will impact the performance of new stores.

It is worth noting that retailers implicitly acknowledge the significance these factors have on store performance in the millions of dollars they spend annually on management, advertising, and merchandising. It is therefore naive to expect that any forecasting system would be capable of accurately assessing every opportunity when so much of a store's performance depends on successful execution on the part of the retailer.

Other factors which contribute to discrepancies between actual and projected sales include road construction activity, changes in the competitive environment, economic downturns, and errors in judgment on the part of the research analyst. In the context of all these factors, it sometimes seems remarkable that there is any semblance between actual and projected sales.

In summary, a sales forecast should be viewed as an estimate of the sales potential present within each site's trade area which has been determined based on the sales performance of existing stores, and whatever eccentricities are inherent in the performance of these stores. These forecasts are more likely to represent actual sales performance when based on large databases, and less likely when based on small databases. But almost always, the correct long-term strategic decision will be made when using sales forecasts as a guide. That is, go/no-go decisions will be accurate, stores will be optimally located to serve customers. We believe that this represents the most important measure of the accuracy of a sales forecasting system.

General Observations

Among the obvious, but easily overlooked, cautions for conducting a thorough site location study is allowance of sufficient time to do the job correctly. It is often not be possible to provide the required answers and sales forecasts immediately. It is certainly inappropriate to expect anything other than a rough estimate of a feasible deployment strategy prior to data collection and fieldwork. The quality of a sales forecast or strategic deployment recommendation is often dependent upon the time and effort that goes into it.

One of the strongest cautions concerning forecasts is illustrated by the unusual experience of a retail chain. The routinely accepted (and intuitively logical) response to the opening of more stores of the same chain in a market was that existing stores will lose sales to new units. That is, some of the market potential for the new stores will come at the expense of existing units. Some of this chain's high-volume, older units, however, which were literally "bursting at the seams," experienced an unexpectedly low sales transfer when nearby stores were opened. After investigating this issue, the retailer found that the reason was that the overcrowding of the older facilities (prior to the opening of the new unit) was such that many potential customers were effectively being "driven away," with the result that the stores were, by virtue of their own success, incapable of achieving their full sales potential. Once the chain added new stores, the initial "stay away" customers were now more likely to come in, since the older units were no longer as busy, and were therefore able to provide acceptable levels of customer service. This phenomenon is unusual, but illustrates a unique situation that can be uncovered by careful analysis.

This observation is a lesson that openness and avoidance of bias toward preconceived and routine patterns and norms is essential for the analyst. In general, the anecdote reinforces the golden analytical rule: Be cautious in undertaking the analysis, but do not let cautiousness (or traditional patterns) blind your judgment to the point of biasing the findings in favor of easily and intuitively acceptable norms.

12 Using Consumer Research to Benefit Site Selection Strategies

RECENTLY, A RETAILER WITH A VERY UPSCALE APPEAL completed the development of a site evaluation database. This retailer then lamented over the findings derived from the process, not because they were believed to be inaccurate, but rather, because of their implications for new store opportunities. More specifically, analysis revealed that our client's future stores should be located in trade areas which encompass large concentrations of very high-income consumers. Needless to say, the fact that there are a limited number of such trade areas nationwide suggests that expansion opportunities for this retail concept as it exists and operates today are also limited.

In the context of these findings, the retailer must now decide whether he or she can be content with its current concept, which has such a narrow consumer appeal that it inhibits its ability to grow, or whether he or she should modify the concept in order to enhance its growth potential. The retailer is faced with an extreme example of a dilemma that is common to most retailers; what changes could and should be made in order to enhance its performance. Of equal concern is will such changes alienate the concept's current base of core customers, thereby resulting in diminishing rather than improving performance. Further, should the concept seek to enhance its market share among its current customer base (e.g., high-income consumers); should it attempt to expand its base of customers by developing a broader appeal; or should it attempt to do both?

As the preceding example demonstrates, these types of issues, which can have significant implications on the site selection process, are generally resolved through consumer research. For the purpose of this book, we are defining consumer research to mean the process (or processes) by which consumers are asked critically to assess a retailer (or retailers) on potentially numerous issues (e.g., operations, merchandising, pricing, service, and so on). This will enable the retailer to develop an action plan to serve its current customer base better, and/or expand.

One of the most basic considerations that should be addressed when designing a consumer research project is which types of consumers should be surveyed. Other factors that should be considered include which method represents the best approach for reaching survey participants (by telephone, in stores, using focus groups), how to obtain a sample comprised of the types of consumers in which the researcher is interested, how many interviews should be conducted among others. These basic issues will be addressed in this chapter. However, as the focus of this book is retail site selection, we do not believe that it is appropriate for us to present a detailed discussion of the various approaches to consumer research; entire books have been dedicated to this subject alone. Rather, the purpose of our discussion regarding consumer research is to inform the reader that these approaches have been used extensively to address real estate–related consumer research issues (as well as a multitude of issues unrelated to the site selection process), and to provide the reader with a basic appreciation for the factors which should be considered when conducting consumer research.

Before embarking on a consumer research study, the researcher must be certain that the purpose of the consumer research project is clearly defined, for it has significant implications regarding how to best proceed on the project. The following list provides examples of some of the more common retail concerns which can be addressed via consumer research:

1. **Us versus them**. How are my stores perceived relative to my competitor's stores; What do consumers like and dislike about not only my stores but also my competitors; and How can I use this information to gain a competitive advantage?

2. **Customer perceptions**. How are the new services, displays, hours, merchandise lines, etc. that I have recently introduced being received by my customers? What do they like about them and what could be done to improve them?

3. **Anticipating reactions**. What will my existing customers think of the sweeping changes I am contemplating in order to expand the breadth of my consumer appeal? Can they live with these changes or will they feel alienated?

Retailers that are attempting to resolve "us versus them" issues are primarily interested in determining how their concept "stacks up" vis-à-vis the competition. By identifying what consumers like about the retailer, as well as its competitors, the retailer is able to accentuate its strengths and improve upon its weaknesses, with the ultimate goal of enhancing its market share. Consumer research studies that have an "us versus them" purpose are typically best addressed via research with consumers at large rather than in-store customer surveys. The reason for this is that when customers are interviewed in a retailer's store, their opinions are implicitly biased in favor of the retailer—if they did not like the retailer, they would not be shopping there! Therefore, if the retailer is interested in obtaining objective consumer opinions regarding how its concept compares with its competitors, interviews with consumers at large are the most appropriate.

Generally, the most efficient means of reaching consumers at large is through telephone surveys. These surveys are typically conducted within a predefined survey area among potential customers. Defining the extent of the survey area is an important first step as it has implications not only for the usefulness of the information that is ultimately derived from the survey, but also, on a more practical level, how the survey sample is assembled. As with all issues pertaining to consumer research, the researcher should consider who he or she wants to survey and what information he or she wants to glean from these consumers prior to defining a survey area. For example, if one of the goals of the survey is to determine how a retailer compares with its competition, then the survey area must be large enough to encompass several of these competitors; as one can imagine, conducting surveys in an area that is exclusively served by the subject retailer will bias the results in favor of the subject retailer.

Once the survey area is defined, the process of assembling the consumer survey sample begins. The initial goal of this process is to ensure that the sample contains the telephone numbers of consumers that reside within the defined survey area. The least expensive (but least desirable) means of assembling a survey sample is to use telephone books as the source for the sample and to focus only on those residents who live in particular towns or whose telephone number begins with a particular exchange. However, this approach can create problems on several levels. Using telephone books implicitly excludes consumers with unlisted numbers from the sample, which likely introduces biases. Further, it is unlikely that the ideal survey area and the areas served by the various telephone exchanges will be coincidental. Finally, the logistics associated with using phone books (or similar publications which provide address and telephone information) to screen for specific exchanges can quickly become unwieldy.

A more practical alternative is to use one of the numerous vendors that specialize in assembling consumer survey samples. These vendors are able to provide a randomly generated sample of telephone numbers (unlisted as well as listed) for virtually any area of the country. Typically, vendors require that the researcher provide them with a map on which the survey area has been defined or a list of ZIP Codes, census tracts, or counties which comprise the survey area. From this map or list, the vendor is able to provide a list of telephone numbers for consumers residing within the defined survey area.

It is important to ensure that the surveys will be conducted among appropriate segments of the consumer base. For example, the "upscale" client discussed previously would likely want to focus on upper-income consumers, whereas an auto parts retailer would likely want to survey only those consumers who own vehicles. Vendors can refine the samples they are able to provide down to specific segments of the consumer pool (e.g., consumers by occupation, income, or age), although generally, the more narrowly defined the consumer segment, the less representative the sample is. This is because often times vendors are dependent on self-administered questionnaires and information obtained from product warranty cards for information pertaining to narrow segments of the consumer base (e.g., consumers that own personal computers). A relatively simple way to ensure that surveys are conducted with appropriate consumers is to obtain a random sample for the survey area, but screen each respondent before the questionnaire is administered. For example, the initial question an auto parts retailer may wish to ask all survey participants is whether they own a vehicle. Those who would complete the survey, while the survey would be terminated among consumers who do not own vehicles.

While some retailers choose to conduct the actual telephone interviews themselves, the logistics associated with this process can be overwhelming. We encourage retailers that are contemplating conducting telephone surveys to consider hiring an agency that specializes in the telephone interviewing process. Such agencies have "banks" of phone lines, and are usually able to efficiently conduct the requisite interviews. Further, these agencies can provide help in refining a questionnaire, and in processing the raw survey data into a format that is easily interpretable.

When conducting telephone surveys, it is important to determine how many interviews are required in order to provide results within a desired accuracy level, as the cost to conduct a single telephone interview is not insignificant. Unfortunately, the optimal sample size cannot be generalized as it is a function of a multitude of issues such as the accuracy of the results required by the retailer, the size of the "universe" of potential survey respondents, budgetary considerations, and the like. Conservatively, a random sample of 300 interviews will provide accuracy within acceptable levels for most retailers. However, retailers are encouraged to investigate the requisite sample size as it relates to their specific research project. When the universe of respondents is small, the sample size can be reduced; conversely, the sample size may need to be expanded beyond 300 surveys if the results are to be prepared for various sub-components of the total sample (e.g., results presented for high-income, middle-income, and low-income respondents). Obviously, it is necessary to have a greater number of potential respondents available for calling than the necessary sample size, since there will always be some refusals and some potential respondents that are not available despite repeated call-backs.

While the preparation described above is essential to the execution of a successful telephone survey, by far the factor that most significantly contributes to the success of <u>any</u> consumer research study is the design of an appropriate questionnaire. Successful questionnaire design blends art and science, as well as experience and intuition. Further, each questionnaire that is designed is as unique as the issues it is intended to resolve. Because of this, it is not practical (nor is it appropriate) to provide the reader with a step-by-step approach to questionnaire design within a text whose expressed focus is retail site selection. We do offer the following tips which have served us well in the development of questionnaires for our clients:

- Be aware of the customer base with whom you are communicating. Many consumers are unfamiliar with terminology that is uniquely retail-oriented (e.g., dark hours, SKUs, price points, etc.). As a general rule, the most successfully executed questionnaires are the ones that are the easiest to answer.

- Within practical considerations, shortest questionnaires are the best. Respondents quickly lose patience; do not waste time asking for information that would be "nice to know" or you may not get the information that is "critical to know."

- Avoid inadvertently prejudicing the opinions of consumers. It is generally best to avoid asking respondents to chose from a list as this inherently limits their choices; let respondents tell you who has, for example, the lowest prices. Also, in most (but not all) cases, it is best to maintain the anonymity of the sponsor of the consumer research.

- Avoid leading questions. Rather than asking "Does Retailer A have good prices?" it would be better to ask "How would you rate prices at Retailer A?"

- Carefully outline all of the issues which will be addressed in the questionnaire and organize them into common groupings. It is frustrating for consumers to "bounce around" from topic to topic.

Examples of exit and telephone interview surveys are provided in Figures 12-1 and 12-2. They have been presented to provide an indication of the types of issues routinely addressed with consumer research. However, we caution the reader against using these questionnaires verbatim. As we have already stated, questionnaires should be designed to address the unique concerns of each retailer and moreover, a well-conceived questionnaire is the most important contributor to successful consumer research.

figure 12—1

SUMPTER CENTER
EXIT SURVEY
FINAL (Date)

LOCATION	1	2	3	4	5	6	7
DAY	**1** MON	**2** TUES	**3** WED	**4** THUR	**5** FRI	**6** SAT	**7** SUN
TIME	**1** 10:30–12	**2** 12–2	**3** 2–4	**4** 4–6	**5** 6–8	**6** 8–9	

INTRODUCTION: Hello, my name is ___________ from ___________. I am taking a survey for Sumpter Center, which will help us serve you better.

Q 1. Have you completed your shopping trip at Sumpter Center today?

1. Yes **(CONTINUE)**
2. No **(THANK & TERMINATE INTERVIEW)**

Q 2. How long was your shopping visit at Sumpter Center today? **(ENTER LENGTH OF VISIT IN SPACES PROVIDED BELOW. EXAMPLES: 1 HOUR AND 15 MINUTES; 2 HOURS AND 0 MINUTES, ETC.)**

_______ hour(s) and ________ minutes

EXIT SURVEY (Continued)

Q3./Q4. I am going to show you a list of the stores, services, and restaurants in Sumpter Center. Please tell me which of these stores you visited today, regardless of whether or not you made a purchase. Also, please tell me approximately how much you spent at each store. **(SHOW CARD; RECORD A ZERO IF RESPONDENT VISITED A STORE BUT DID NOT MAKE A PURCHASE; ASK RESPONDENT FOR CORRESPONDING CODE #; CIRCLE AND FILL IN AMOUNT)**

1. $ .00
2. $ .00
3. $ .00
4. $ .00
5. $ .00
6. $ .00
7. $ .00
8. $ .00
9. $ .00
10. $ .00
11. $ .00
12. $ .00
13. $ .00
14. $ .00
15. $ .00
16. $ .00
17. $ .00
18. $ .00
19., $ .00
20. $ .00
21. $ .00
22. $ .00
23. $ .00
24. $ .00
25. $ .00
26. $ .00
27. $ .00
28. $ .00
29. $ .00
30. $ .00
31. $ .00
32. $ .00
33. $ .00
34. $ .00
35. $ .00
36. $ .00
37. $ .00
38. $ .00
39. $ .00
40. $ .00
41. $ .00
42. $ .00
43. $ .00
44. $ .00
45. $ .00
46. $ .00
47. $ .00
48. $ .00
49. $ .00
50. $ .00
51. $ .00
52. $ .00
53. $ .00
54. $ .00
55. $ .00
56. $ .00
57. $ .00
58. $ .00
59. $ .00
60. $ .00
61. $ .00
62. $ .00

( ) Other:____________________ $ .00

( ) Other:____________________ $ .00

EXIT SURVEY (Continued)

Q5. Did you use the mall's customer service center today?

1. Yes 2. No

Q5A. What services does the mall's customer service center offer? **(DO NOT READ LIST; ALLOW MULTIPLE RESPONSES; PROBE BY ASKING "ANYTHING ELSE?")**

3. Community room rental
5. Copying
7. Event information
8. Faxing
9. Gift certificates
13. Lost and found
15. Mall employment
19. Notary
21. Postage stamps
25. Save & Win Club membership registration
26. Security assistance
27. Shopping bags
30. Store information
31. Strollers/Wheelchairs

( )
( )
( )

Q6. Were there any items that you were shopping for but were unable to locate here today? **(ASK "WHAT ITEMS;" ALLOW MULTIPLE RESPONSES; PROBE BY ASKING "WHAT ELSE?")**

1. No/None

____
____
____

Q7. How many times during the last 3 months have you visited Sumpter Center, including today? **(ENTER A SINGLE NUMBER NOT A RANGE OF NUMBERS. EXAMPLE: 4 TIMES, NOT 3–5 TIMES)**

OF VISITS __________

98 Don't know/Refused

EXIT SURVEY (Continued)

Q8. How would you grade Sumpter Center on the following aspects where A = excellent, B = above average, C = average, D = below average, and E = poor or unsatisfactory? **(READ LIST; ROTATE ORDER; ALLOW ONE RESPONSE PER CATEGORY)**

	A	B	C	D	E	Don't Know
1. Variety/selection of department stores	4	3	2	1	0	8
2. Variety/selection of other stores	4	3	2	1	0	8
3. Variety/selection of restaurants	4	3	2	1	0	8
4. Variety/selection of the food court	4	3	2	1	0	8
5. Nice shopping environment	4	3	2	1	0	8
6. Ease of finding your way around	4	3	2	1	0	8
7. Feeling safe when shopping	4	3	2	1	0	8
8. Reasonable prices	4	3	2	1	0	8
9. Good parking	4	3	2	1	0	8
10. Quality merchandise	4	3	2	1	0	8
11. Service received in stores	4	3	2	1	0	8

EXIT SURVEY (Continued)

Q9. Are you a member of the Sun and Fun Club here at Sumpter Center?

1. Yes 2. No **(SKIP TO QH)**

Q10. Which of the following Sun and Fort Club services do you use? **(READ LIST; ALLOW MULTIPLE RESPONSES)**

1. Coupons
2. Free copy service
3. Free fax service
4. Free shopping bags
5. Free stroller rental
6. Have won a prize
7. Receive the newsletter

These last few questions will help us categorize all the information you gave us.

Q11. Are you a tourist, business visitor, seasonal resident, or year-round resident in the Phoenix area?

1. Tourist
2. Business visitor
3. Seasonal/Winter resident
4. Year-round resident

( ) Other: ____________________

6. Refused

Q12. How long have you lived at your present address? **(READ LIST; ALLOW ONE RESPONSE)**

1. Less than 1 year
2. 1 to under 3 years
3. 3 to 5 years
4. More than 5 years
5. Refused

Q13. What is your local home ZIP Code?

_ _ _ _ _ **(ENTER 99999 IF DON'T KNOW/REFUSED)**

EXIT SURVEY (Continued)

Q14. Which of the following best describes your marital status? **(READ LIST)**

1. Single
2. Married
3. Widowed
4. Divorced
5. Separated
6. Refused **(DO NOT READ)**

Q15. How many people including yourself reside in your household? **(DO NOT READ)**

1. One
2. Two
3. Three
4. Four
5. Five
6. Six
7. Seven or more
9. Refused

Q16. How many children under the age of 10 currently reside in your household? **(DO NOT READ)**

0. None
1. One
2. Two
3. Three
4. Four
5. Five or more
6. Refused

Q17. Which category best describes your age? **(SHOW CARD; ALLOW ONE RESPONSE)**

M 1. Below 20
F 2. 20–29
Q 3. 30–34
H 4. 35–44
Z 5. 45–54
S 6. 55–64
R 7. 65 or older
8. Refused

Q18. Which of these categories best describes the total annual income of everyone in your household? **(SHOW CARD; ALLOW ONE RESPONSE)**

H 1. Up to $14,999
T 2. $15,000 to $24,999
A 3. $25,000 to $34,999
L 4. $35,000 to $49,999
S 5. $50,000 to $74,999
Q 6. $75,000 to $99,999
M 7. $100,000 to $149,999
B 8. $150,000 and over
9. Refused

OBSERVATIONS; DO NOT ASK

Q19. Gender

1. Male
2. Female

Q20. Race

1. Caucasian/White
2. African-American/Black
3. American Indian
4. Asian
5. Hispanic
6. Other:

figure 12—2

TELEPHONE SURVEY
Final (Date)

Survey # ________________ Time ________________________________
Date ______________________ Interviewer Initials ____________________

Introduction

Good evening, my name is ______________________ with _____________ and we are conducting a survey about shopping habits. This is not a sales call.

QA. May I speak with the person in your household that does most of the shopping for clothing, gifts, and things for your home? **(IF NOT AVAILABLE, ARRANGE CALL-BACK; REPEAT INTRO IF NECESSARY)**

1. Yes **(Continue)**
2. No **(Thank & Terminate)**
3. Not available **(Schedule a Call-Back)**
4. Refused **(Terminate)**

QB. What is your ZIP Code? ________________

QC. How many times in the past 3 months have you shopped each of the following malls or shopping areas? **(READ LIST)**

	Frequency in 3 Mos.	Don't Know
1. Salem Mall	_________________	99
2. Bridgeport Mall	_________________	99
3. Danbury Mall	_________________	99

While telephone surveys usually represent the best vehicle for addressing "us versus them" issues, there are some inherent shortcomings associated with this approach. The most basic of these shortcomings is that telephone survey respondents may not have "fresh" impressions regarding the retail concept(s) being evaluated; several weeks may have passed since their last visit to the retailers in question. Moreover, the increasing number of telemarketing calls to consumers' homes have resulted in a commensurate increase in the number of consumers that refuse to participate in telephone surveys; this, in turn, has resulted in escalating costs associated with conducting telephone surveys. Further, telephone surveys are not an efficient form of consumer research when the retailer is interested in assessing the opinions of its customers exclusively; in-store surveys (discussed below) represent the most effective means of conducting research among customers.

To this point, our discussion has focused on retailers that are interested in addressing "us versus them" issues. But what about retailers that are more interested in ascertaining consumer perceptions regarding what is going on <u>inside</u> their stores? Specifically, it is often helpful for retailers to assess how customers perceive their service levels, displays, merchandise quality, selection, and newly instituted programs and policies, or more simply, to determine what customers like and dislike about shopping their stores, and what they would recommend to improve them. These issues are usually best addressed by conducting in-store interviews (as opposed to telephone interviews) with customers.

In-store interviews an are excellent way to assess how a retailer's customers generally perceive their stores, and can also be structured to provide insight into what customers think about a specific aspects of their stores. The reason for this is that unlike telephone surveys, the customer's shopping experience is fresh in their minds when they are interviewed during a visit to a store. In-store interviews can provide the strongest direct reaction to the cleanliness of a facility, its attractiveness, and the customer service just received (or not satisfactorily received).

A disadvantage of in-store surveys is often the pressure of time; customers are usually anxious to be on their way. This pressure limits the practical length of in-store surveys. Further, in-store interviews do not represent an effective means of gathering information regarding customer attitudes towards competitors. The fact that customers have chosen to shop a retailer's store as opposed to one of its competitor's stores is an implicit endorsement of the retailer versus its competitors. Thus, any information collected regarding the retailer versus its competitors will be strongly biased in favor of the retailer.

The process for conducting in-store interviews is generally more straightforward than conducting telephone interviews, primarily because unlike telephone surveys, a survey area does not have to be defined, and a sample of consumer telephone numbers does not have to be obtained; the sample "pool" is the customers that are shopping the stores. However, conducting in-store interviews is not as simple as engaging a few customers in conversation—there are several guidelines that should be followed to ensure that the findings derived from the survey are as representative of the total customer base as is possible.

The most important of these guidelines is that the interviews should be conducted so that they are representative of a store's sales distribution throughout the day, and by day of week. The composition of a retailer's customer base can change dramatically overtime. An example of this is the change that occurs in the composition of the customer base of movie theater patrons throughout the course of a day. The opinions of afternoon movie-goers are primarily those of preteens, whereas the opinions collected during the evening would be those of adults.

To ensure that in-store interviews are truly representative of a retailer's customer base, we recommend that a matrix such as that presented as Table 12-3 be developed for each store in which interviews will be conducted. More specifically, the matrix is designed to indicate the proportion of a store's sales that are realized by hour for each day of the week. Ideally, the surveys should be conducted in accordance with this matrix such that if 4 percent of a store's business occurs during the morning hours on Thursday, 4 percent of the interviews are conducted during this period. As a practical matter, retailers may relax these standards somewhat; for example, retailers may not wish to interview at all during very slow periods, or during the "softest" sales days of the week. The researcher must try to balance the theoretical goal of ensuring the representativeness of the sample against practical concerns such as attempting to collect customer interviews during periods when customer traffic is very light.

TABLE 12-3
INTERVIEW MATRIX

	Mall Sales Distribution	Interview Quotas	Number of Completed Interviews	Interview Days
Weekdays				
(Monday–Friday)	45%	180	90	Thursday
			90	Friday
Weekends				
(Saturday–Sunday)	55%	220	110	Saturday
			110	Sunday
TOTAL	100%	400	400	

The time of year should also be a consideration when conducting in-store surveys. For example, a disproportionate share of toy-store sales are likely generated during the Christmas season, whereas the months following Christmas likely represent their slowest sales periods. In this context, surveys conducted in November and December would yield results that are more representative of reality relative to surveys conducted in February.

Generally, in-store interviews should be conducted with customers who have made a purchase rather than browsers. This is because browsers may not have any intention of ever making a purchase in the store, and therefore, their opinions should not be considered when developing a future game plan for the store. Consumers who like to browse at luxury car dealerships but do not have the financial resources to make a purchase represent an extreme example of this concern. It should be noted that there are instances when it is appropriate to interview nonpurchasers, such as to identify why a purchase was not made. As with all consumer research, the goal of the research must be clearly understood prior to embarking on the research project. If the primary goal is to focus on the opinions of customers, do not interview browsers; conversely, if the primary goal is to understand why browsers are not making purchases, then browsers should be interviewed (although identifying the reasons that consumers do not shop a store may be more efficiently handled via telephone surveys).

When conducting in-store surveys, interviewers should generally be stationed at points within a store where customers enter and exit. By doing so, the researcher ensures that a representative sample of customers are interviewed. As an example, if interviewers are positioned in the footwear department of a sporting goods store, they will capture a disproportionate share of footwear shoppers, but an underrepresentation of shoppers in other departments. Retailers with drive-through facilities (such as quick service restaurants and banks) should not overlook these areas; interviewers should be positioned to survey these customers as well as customers within the store.

Finally, interviewers must not be allowed to introduce sample bias by being selective as to the types of customers they interview (e.g., interviewing only well-dressed as opposed to "grubby-looking" customers). Generally, interviewers should be instructed to address every Xth customer regardless of their appearance, sex race, age, and the like.

It is important to remember that customers often perceive the interviewers as representatives of the store. Accordingly, the interviewers should be pleasant with respect to their appearance as well as their demeanor. In-store interviews can be administered by store personnel, but often (particularly among retailers unfamiliar with the consumer research process) it is valuable to hire a professional interview agency. Many of these agencies can provide good advice regarding how the actual survey process should be administered. Further, these agencies are often able to provide tips in the design of questionnaires, and assistance in the processing of the raw survey information into a usable format. We caution, however, that interview agencies are in the business of conducting interviews for a multitude of industries as well as for retailers; they do not necessarily understand the retail industry nor the dynamics of what makes a retailer successful.

In addition to the issues discussed above, retailers will often use in-store interviews to find out more about their customers (e.g., sex, age, visiting from home or during their lunch hour, their demographic make-up, and so on). It is critical however, for a retailer to recognize that the information obtained via this process merely reflects the demographics of persons shopping the store as opposed to the retailer's "customer profile," which refers to the specific and unique demographic segment of the consumer base to which the retail concept most strongly appeals. The demographics of a store's shoppers are largely self-fulfilling; that is, they are directly a function of the types of consumers that reside within the store's trade area. For example, if The Gap were located in a retirement community, most of its shoppers would be elderly. Similarly, if MacFrugal's were located in an affluent community, most of its shoppers would be affluent. In both instances, it would be erroneous to conclude that the demographics of the customers shopping these stores are, in any way, related to the retailer's customer profile—that is, the types of consumers to which these retail concepts most strongly appeal. Rather, a retailer's customer profile is best determined by identifying the demographic characteristics of consumers who are most inclined to shop the retailer.

As with telephone surveys, the most important contributor to the successful execution of in-store interviews is the design of a concise and well-conceived questionnaire. The reader is strongly encouraged to consider the commentary presented earlier in this chapter when designing a questionnaire to be administered within his or her stores.

Determining how many interviews should be conducted for in-store consumer research is as nebulous as it is for telephone surveys. Unlike telephone surveys, in-store surveys are relatively inexpensive to administer and therefore, as a general rule, the larger the sample size, the better. However, there are practical limits to how large an in-store survey should be; unfortunately, as with telephone surveys, the optimal sample size for in-store surveys varies depending on the unique scope and nature of each study. As stated previously, a sample of 300 interviews will generally provide accuracy within acceptable tolerance levels for most retailers, although the requisite sample size may need to be larger or smaller depending on the specific requirements of each consumer research project.

Based on the results of consumer research and/or intuition, retailers will often propose broad or sweeping changes (relating to operations, advertising, merchandising, etc.) which are intended to enhance the appeal of their concept to existing customers, or to expand their concept's appeal to a broader array of consumers. It is worthwhile to attempt to gauge how consumers and customers will react to these changes prior to their implementation.

Anticipating consumer reactions to change generally necessitates a research format that is less structured than in-store or telephone surveys. For example, if a retailer is contemplating changes to its merchandise selection, it is not enough to know that consumers simply do not like the store's current selection—much more specific information is required. What specifically does the consumer not like about the selection? Are there aspects of the selection that the consumer does like? Is it really the selection that the consumer does not like, or is it, in fact, the presentation of the merchandise? These are but a few of the issues a retailer contemplating changes to its merchandise selection needs to resolve.

Even the most organized and efficiently designed questionnaires cannot provide for all the issues that may arise when a retailer is attempting to anticipate consumer reaction to change. Thus, when broad, conceptual issues are to be investigated, focus groups (rather than their more structured telephone and in-store interview counterparts) represent an appropriate form of consumer research.

Simply stated, focus groups involve assembling a group (or groups) of consumers to discuss issues of concern to a retailer for the purpose of providing insight into the opinions and perspectives of the consumers. Typically, the size of each group ranges between ten and fourteen participants, and usually, at least two groups are conducted. As stated previously, the strength of focus groups is that they are "free-form" and thereby are flexible enough to explore relevant tangents. The single biggest weakness of focus groups is that they are based on a very small sample of consumers. Due to the qualitative nature of focus groups, the conclusions derived from them, while providing a reasonable indication of consumer sentiment, are not "statistically reliable," as are the results of quantitative research (e.g., telephone and in-store surveys).

As with all consumer research, the first step in conducting focus groups is to identify clearly the issues to be addressed, and to determine what the ultimate goal of the research should be. This step has implications for recruiting focus group participants as the researcher must understand what needs to be resolved before it is possible to identify the type of individuals that should be recruited as focus group participants. For example, if the goal is to enhance a store's market share among its current customer base, participants should be current customers. Alternatively, if the goal is to expand the appeal of the retail concept, then participants should not be customers (instead, they could be "regular shoppers" of a competitor[s]).

There are numerous ways to recruit focus group participants. If participants are to be customers, then they can be recruited from customer lists, credit card lists, and so forth. Further, as was discussed during the portion of this chapter devoted to telephone surveys, a list of consumers residing within a defined area can be obtained from various vendors. Using this list, a simple questionnaire can be developed for the purpose of screening for appropriate focus group participant candidates (e.g., have you shopped at Store X within the past month). While focus group participants are generally recruited via telephone, it is also possible to recruit in conjunction with in-store surveys (i.e., obtain names and phone numbers of customers who were shopping store X during a particular time frame). We generally recommend that candidates are screened to ensure that store employees and persons who have participated in a focus group within the prior six months are not included in the groups. Figure 12-4 presents a sample focus group screening questionnaire.

figure 12—4

FOCUS GROUP SCREENER
Final (Date)

Introduction

Hello, my name is ____________________ with ______________________ and we are conducting a very short survey about shopping habits in **(area)**. This is not a sales call.

Q1. Have you or any member of your household shopped for appliances for your home within the past year or so?

1. Yes **(Ask "May I speak with the person that was primarily responsible for deciding where to make your appliance purchase?") (IF NECESSARY, ARRANGE A CALL–BACK. REPEAT INTRO)**
2. No **(Thank & Terminate)**
3. Don't know **(Terminate)**

Q2. At which one store did you purchase the appliance?

____

Q3. I would now like to ask you about your shopping habits for home improvement items. Which of the following stores have you shopped at least two times within the past year or so for home improvement items? **(READ LIST; ALLOW MULTIPLE RESPONSES)**

1. Kmart
2. Wal-Mart
3. Builders Square
4. Sears
5. Home Depot
6. Lowe's

It is usually necessary to provide participant candidates with a monetary incentive to participate in the groups. Further, it is usually necessary to hold the focus groups during evening hours so that it is convenient for the participants. Most retailers that conduct focus groups elect to work with a consumer research agency that specializes in focus groups and has facilities designed to accommodate these sessions. Usually, a focus group facility consists of a meeting room in which the participants and the focus group moderator address the various issues, and a viewing room in which the retailer may observe the focus group sessions; a one-way mirror divides the two rooms.

The agency can be of considerable assistance in facilitating the focus group process. They will arrange for appropriate candidates to participate in the focus groups, arrange for audio and/or video taping of the groups, coordinate payment of participants, and arrange for refreshments for participants and the client. Many of these groups will also provide a focus group moderator, assist in the development of a focus group discussion outline, and provide a summary of the findings of the groups. However, as with the other agencies discussed herein, these agencies are not necessarily retail experts; it is our belief that a good moderator should have considerable moderating experience and a strong background in retailing in order to achieve the most meaningful results from focus groups.

Just as questionnaire design is the most important step in conducting quantitative consumer research (telephone and in-store surveys), the design of a focus group discussion outline is the most important step relating to successful focus groups. All of the cautions addressed earlier in this chapter regarding questionnaire design apply to the development of a well-conceived focus group discussion outline.

Consumer research has proven to be helpful in identifying and measuring the reaction of customers and noncustomers to a retailer's facilities, merchandising practices, and operational practices. It can provide evidence for relative strengths or weaknesses in a retail facility and, as noted, can assist in maximizing the sales performance of any existing store. The consumer research tools discussed in this chapter by no means represent all of the possible approaches used by retailers to provide insight into the opinions and perspectives of consumers. However, they are among the most commonly used and proven approaches.

13 Start-Up Stores

START-UP STORES ARE NOVEL, INNOVATIVE RETAIL FACILITIES that present a dramatic change in the traditional way of doing business and presenting merchandise. Start-up stores distinguish themselves from existing retail operations in:

- format (especially scale—e.g., superstores)
- merchandise selection (offering, e.g., a unique combination of goods, such as wine and cigars)
- specialization (catering to a distinct end-user group, such as infants)
- image (new or innovative layouts)
- pricing (warehouse clubs, for example)
- customer appeal (combining sales to groups that were previously addressed separately)

Compared with more mature retail concepts, unique start-ups have the appealing prospect of rapid initial growth, thus, they are obvious targets for attracting venture capital. Much more than any other retail development, start-ups offer the attraction of great financial reward, but are usually also associated with a commensurate level of risk.

There are two types of start-up stores: (1) an industry start-up that features entirely new ideas in store operations, merchandising, or merchandise groupings; and (2) "clones" of original start-up operators. Toys "R" Us is an example of an industry start-up, as are The Home Depot and Kmart. While clones are also start-ups, they do not establish a new retail precedent, rather, they mimic (and sometimes refine) the ideas and innovations of the original start-up. For example, The Home Depot was the first true warehouse-format home improvement center; Home Quarters and Builder's Square are clones of The Home Depot.

The Location Research Challenges for Start-Ups

There are logical, definitive methodologies for multistore retailers to create databases and use them for sales forecasting and strategy analysis purposes. However, for start-up stores, and particularly the industry start-ups, there is no internal data that can be used in database development, as there are no comparable stores. Not only is there no database for sales forecasting purposes but much of the elementary information that results from database development (e.g., the geographic extent of store trade areas, the identification of competitive impacts on store performance, the demographic characteristics most strongly associated with favorable sales performance) is unavailable. Despite these obvious shortcomings, the location research principles of retail site selection are still applicable for start-ups even though the usual background information on which they depend is not available. The research analyst who has the task of identifying site opportunities and forecasting sales for the industry start-up must recognize that despite its uniqueness, the concept is still subject to the locational rules and guidelines which apply to all other retailers. The start-up must locate at a site with access to sufficient densities of demographically in-profile customers for its concept.

The challenge to the researcher, then, is to generate as much background information as possible about the concept and develop that information into usable sales forecasts. Indeed, the researcher must be creative in the development of alternative approaches to generating background information, and use that information to develop indirect methods and location inferences.

Compared to the industry start-ups, database development for clones can be based on more directly observable data. Specifically, clones can draw upon the experiences of their competition; that is, they can use the successful industry start-up as the basis for the development of an appropriate site selection strategy. The major disadvantage for clones is that they are playing "catch-up" and may not be able to overcome the dominance of the industry start-up concept. Nonetheless, many clones have done very well, particularly if they learn well from the original start-up, even moving on to a position of industry leadership. A good example of this is The Sports Authority, which was patterned in part on the now defunct Total Sports and All American Sports Club.

Consumer Research for Start-Up Stores

Prior to considering a location deployment strategy for a start-up store, it is important to determine consumer reaction to the concept; in the opinion of consumers, does the concept fill a consumer need and generate enough consumer interest and excitement to suggest that consumers would be willing to shop the new concept? With start-ups that are totally new concepts (i.e., the industry start-up), consumer reaction to them can be tested in several ways. Further, critical information on where the concept's potential customers are currently shopping for the goods that the start-up will offer, and what the lure might be to get them to go to different formats and locations can be readily determined through consumer research.

One way to address the potential prospects for a start-up is to conduct focus group research. If multiple focus groups are conducted, the information gleaned from the earlier groups can be used to "fine-tune" the direction of questioning in the later groups. As an example, participants in an early focus group might indicate that they shop for the goods to be offered by the start-up at a store that had not previously been considered to be a competitor. The later focus groups can be probed to determine their likes and dislikes about this newly identified competitor, with the goal of ensuring that the start-up concept makes appropriate merchandising and operations adjustments to enable it to capitalize on the competitor's weaknesses.

Focus groups will be made up of a moderator equipped with a script on the topics for the group to address, and 10 to 14 persons who have been screened to ensure that they are potential patrons of the start-up. Ideally, a focus group session should have a mock-up of the store or concept being tested. If merchandising and store layout are part of the focus group agenda, visuals demonstrating the concept might be used. These visuals may include architectural drawings of both the interior and exterior of the store, as well as examples or videos of the merchandise that the start-up will carry. A detailed explanation of how the store will operate, the merchandise it will carry, how it will be displayed and priced, and other pertinent information should accompany the visuals.

Conducting several focus groups to explain the start-up concept and gathering opinions (both positive and negative) greatly benefits the start-up retailer. Focus groups allow the retailer to refine its concept so that it addresses the needs and preferences of the targeted consumer. Further, focus groups provide insight about how the concept will be accepted, if it will be shopped more or less than its conventional competitors, where it should be located, the hours it should keep, and who its customers are likely to be.

If the insight gained from the focus groups is inadequate or inconclusive, or there is disparity in the opinions given by focus group participants that cannot be resolved, it will be necessary to generate additional information. In most instances, the information can be obtained from telephone interviews or other types of consumer research. For example, before Staples, the large-format office supply stores their used supermarket approach to merchandising office supplies, it made sense to answer several questions regarding office supplies purchasing. Who makes office supplies purchase decisions? What is the importance of delivery? What is the importance of selection? How important are everyday low prices versus special sales? Consumer research was used to not only address these specific questions, but also to explore the potential associated with nontraditional buyers of office supplies such as schools or individuals who have work-at-home businesses. In general, consumer research is used to obtain the basic parameters concerning the sales potential for the start-up retail store.

Well thought-out consumer research is generally helpful in defining what the new start-up should be, and what factors will most favorably distinguish it from its traditional competitive alternatives. However, there are certain aspects of store performance that even the most well-conceived consumer research cannot anticipate. Returning to the example of office supply stores, no consumer research study would have predicted how significant the sale of school supplies, office supplies to individuals, and home office equipment would be in the context of overall store performance, since these were market segments not well served by the traditional office supply segment. Nonetheless, the consumer research would likely have indicated that the office supply superstore concept would be able to realize this potential, but not quantify its extent. Thus, consumer research can help to refine the scale of an individual store and give an indication as to the possible upside limit of its business. On a more global basis, consumer research can be structured in such a way so as to indicate the number of units that ultimately could be supported. This can be determined by asking respondents how much they might spend annually at the start-up concept, how far they would be willing to travel to visit the store, and how often they might shop.

While consumer research can provide an indication of market acceptance and maximum store build-out, it offers little direct insight into the most appropriate deployment of individual stores or the development of market location strategies. Consumer research information can be helpful in determining where consumers are now shopping for merchandise that will be offered by the start-up, from which inferences regarding appropriate locations can be derived. As an example, it would be a significant finding if consumer research indicated that consumers in a market would be willing to travel a significant distance in order to benefit from the low prices and greater variety offered by a pet supply superstore. Consumer's willingness to travel a significant distance indicates that the location consideration should be that of the destination-oriented retailer. The retailer now knows that it should be looking for sites with accessibility to a large area. Also, because their concept is destination-oriented, they can be less concerned about cotenancies in shopping centers. Finally, consumer research may be beneficial in demographically orienting the site selection process. For instance, if the consumer research suggests that the concept will have a greater likelihood of success with upper-middle income consumers, at least for the first wave of stores, the analyst will want to screen the market for upper-middle income areas with sufficient population densities. Thus, intelligent use of consumer research allows the retailer to concentrate on locations in demographically "in-profile" areas.

Using Available Industry Data

The information obtained from focus groups or telephone surveys is only a part of the background data necessary to develop guidelines that are useful in undertaking location analysis for start-up retail stores. While consumer research can help to refine the concept, provide an indication of its acceptance, and infer the optimal locational characteristics, it cannot define the sales potential for the concept or any other quantifiable data that would be relevant to the sales forecasting process. Data obtainable from government agencies, however, can be helpful in assessing the viability of start-up concepts. For instance, retail sales information by various merchandise categories (food, sporting goods, clothing), as well as by retail outlet types (food stores, sporting goods stores, department stores) is generally available either from Census data or other sources such as retail sales tax collection figures for individual states. These figures are a gauge of market volume potential for various merchandise lines or combinations of lines. This data, in turn, may be broken down into sales volume potentials for regions of the country, states, metropolitan areas, or other market segments. By making assumptions regarding the overall market share that could be obtained by the start-up, it is possible to estimate an aggregate volume that the retailer can generate from the market, and the total number of supportable stores.

As an example, if one million people in a metropolitan area each was estimated by the U.S. Department of Commerce to spend $250 per year for items typically found in sporting goods stores, the market would have a sales total potential of $250 million annually. If the start-up was a large sporting goods store format and was assumed to be able to capture a 25 percent share of this market, it could generate approximately $62.5 million in annual retail sales from this market. Further, if each of its stores was assumed to average $9 million in annual sales, it could develop up to seven stores in the market. The information available to calculate sales potential and the assumptions regarding market share could be used to evaluate a larger area such as a region of the country or the country as a whole. Further, the analysis of this information can be especially helpful for financial backers who are seeking to gauge the ultimate size of a start-up concept. Using this information, generalities can be made as to the total number of potential units that could be developed, and the approximate time it would take for the start-up concept to reach saturation (maximum build-out). In turn, this information would be helpful in determining when a public offering might have maximum benefit.

Generating Data for Start-Up Clones

While companies attempting to clone a retail start-up also have consumer research and secondary data available to them, they are often able to assess critical location-related issues more accurately (such as the trade area extent and customer profile of future stores) by conducting research on stores operated by the original start-up company. Ideally, the clone company would conduct intercept surveys with customers who are shopping the start-up stores. This approach is not feasible as no retailer would knowingly allow a potential competitor to interview its customers. A more likely way of obtaining information on this store would be to conduct a license plate survey among its customers. For a relatively modest fee per vehicle, many states will provide vehicle registration addresses which correspond with vehicle license plate numbers.

In this system, observers are positioned in the parking lots of the subject retail stores. The data collectors observe shoppers as they leave the store to determine if they had made a purchase at the store. Assuming the shopper had made a purchase (by virtue of the fact that they are carrying a bag from the store), the license plate number of customer's vehicle would be noted. Several days of customer observations are generally sufficient to collect a statistically valid sample. The license plate numbers would then be sent to the appropriate state agency which can provide a list of addresses corresponding to the vehicle license plate numbers.

Once the data is received from the state, the addresses can then be geo-coded and a distribution of the store's customers can be mapped. From this map, a primary trade area showing a concentration of about two-thirds to three-fourths of the store's shoppers can be defined. Based on this trade area definition, the analyst can then:

- Calculate the population base residing within the trade area.
- Determine the trade area's demographic characteristics.
- Estimate an area within which telephone interviewing can be conducted for the purpose of determining consumer likes/dislikes regarding the competitor.

By determining the population base in the trade area, the analyst now has a better idea of how much density is required to support the store. Further, by examining the demographic data associated with the trade area, the analyst can determine if the store is getting more or less patronage out of certain demographic areas. As an example, if, on a per-capita basis, the store is getting twice as many shoppers from very-high-income areas than from middle-income areas that are approximately the same distance from the store, the analyst has at least some insight regarding the customer profile of the store to be cloned.

With trade area parameters defined for the store to be cloned, the analyst now has a geographic area within which to conduct consumer research. Specifically, the analyst should examine the portions of the trade area with the greatest concentration of customer residences. This is the area where the telephone survey should be conducted in order to gather the greatest sample of potential shoppers. Some of the issues to be addressed during the telephone survey will elicit consumer response to:

- reasons for shopping/not shopping the store
- the store's strengths/weaknesses
- how often they shop the store
- how much they spend at the store
- how long they have shopped at the store
- other stores they shop for the same type of merchandise

The consumer research, similar to the population and demographic data, will give the researcher more insight into the store, thus helping in the location research effort for the clone store. It should be clearly noted, however, that the information ultimately gleaned from license plate surveys at stores is imperfect at best. It is usually based on a very limited sample of stores and, obviously, those stores do not share the exact operations, merchandising, or service levels which will ultimately be offered by the clone.

There are some drawbacks with license plate surveys, including the following:

- No purchase amount can be assigned to the license plates; therefore, the store's sales cannot be distributed geographically.

- People move and do not report their address changes right away; this lag results in some inaccurate "matches" between license plate registrations and residences.

- Many vehicles are leased, rented, or supplied to employees; thus, their registrations are matched to an address of a corporation rather than a residence.

Aside from the technical problems related to the accuracy of license plate surveys, there are additional problems related to the cost of conducting license plate surveys. While the cost per vehicle is fairly low (in some states $5 per vehicle or less), the cost to develop a statistically significant sample (typically 400 or more vehicles) can be fairly high. Also, if possible, the survey should be conducted at several stores to minimize the risk of overgeneralizing about factors such as trade area size, customer appeal, and so on;

surveying in multiple stores results in much higher costs. Further, collecting and processing the license plate surveys can be quite time consuming as field personnel have to obtain the listings of license plates, send them out for processing and await the return of the printouts. Finally, because of privacy concerns, several states no longer supply information on license plate numbers.

Sales Forecast Implications

There are a few simple steps to follow when conducting a sales forecast for a start-up store with only a limited background of information:

1. **Define the Trade Area for the Subject Site**—The definition of the trade area will be based primarily on the information from the license plate surveys of existing, comparable stores. If these surveys indicate that a store is drawing from a certain distance (e.g., a 6-mile radius) then this information would serve as a guide in defining the trade area for the new store. If vehicle registration information is not available, a depiction of the trade area will need to be made based more on theory and knowledge of similar retail concepts than objective data. Specifically, the trade area will be defined based on accessibility, deployment of competition, barriers, etc.

2. **Calculate Population Data**—Once the trade area is defined, a population estimate can be readily calculated. This information can be obtained from a demographic profile for a radius around the site, or on a subgeography basis (i.e., census tracts or ZIP Codes) in order to more closely replicate the defined trade area.

3. **Define Potential**—The potential for the subject site will likely be based on retail sales of comparable stores or merchandise line sales of products to be sold in the start-up. As an example, if the store under consideration is a sporting goods retailer, the analyst

should check secondary data for sales information on all stores that sell sporting goods. If available, this information should be for the metropolitan area in which the new store will be operating, although it could be extrapolated from statewide data if necessary. The total amount of sales generated by existing sporting goods stores would approximate the sales potential available, but this figure could be augmented by dollars spent at stores other than sporting goods stores for sporting goods. As an example, the potential could be increased to account for the sale of sporting goods through discount department stores. Conversely, the analyst could add up all of the sales for each of the groups of items that are carried by sporting goods stores (athletic shoes, sports apparel, sporting goods equipment). Again, this information should be assembled on a local level if possible.

Both of these methodologies require that the total sales figure be divided by the appropriate population base (either metropolitan area or state) to determine per-capita potential. This figure is then multiplied by the trade area population base to determine the trade area potential.

As an example, if it has been determined that there are 70 stores in the state selling the same type of products as the start-up, and these stores average $7 million in sales, then the total state potential is about $500 million. If the state has a total population of 5 million persons, then the expenditure potential equates to $100 per person per year. If the defined trade area has a population base of 200,000 persons, then the trade area has a potential of $20 million.

4. **Calculate the Share of Space**—One of the simplest sales forecasting methodologies is referred to

as "share of space/share of market." This methodology can provide only rough guidelines and should only be used when no proprietary forecasting data is available, such as for start-up stores.

As the name implies, the percentage of space that a store has will equal its proportion of the total market. Thus, the analyst will need to calculate the total size of all competitors serving the subject trade area, add to that figure the size of the subject store, and find the percentage that the subject store will have of the new total. As an example, if it is determined that there are currently 120,000 square feet of space shared by competitors selling the items that the start-up will sell, and assuming that the start-up will be 40,000 square feet, then the start-up will have a 25 percent share of space (40,000/(120,000 + 40,000)).

5. **Forecast Mathematics**—Knowing that the subject store will have a 25 percent share of space, the forecast methodology of "share of space/share of market" assumes that it will also generate 25 percent of the market potential. Thus, continuing our sporting goods store example, if it has been determined that there is $20 million in potential within the trade area, the start-up store will capture $5 million of that potential.

The only function left to complete would be estimating "beyond" sales for the subject site. Previously, we have noted that a primary trade area normally accounts for about 65 percent to 80 percent of a store's total sales volume. The analyst will need to decide what the appropriate figure for beyond sales will be. As a guideline, the less population density beyond the trade area, or the more competition beyond the trade area, the lower the proportion of beyond sales will be. Assuming that

> the analyst settles on 20 percent of all sales to be generated from beyond the trade area, the resulting beyond sales figure will be $1,250,000 and the total sales volume will be $6,250,000 ($5,000,000/80% [where 80% equals the proportion of sales achieved within the trade area]).

Although data is limited, a reasonable sales forecast can still be developed for start-up stores based on competitors or secondary data. A store database and customer profile analysis should be developed as soon as practical after the first start-up store is opened. Further, this database and customer profile analysis should be augmented with each ensuing store opening until there are sufficient units offering enough data to forecast every possible scenario. At that point, the start-up retailer is now a solid retail chain and its forecasting system should be built as such.

14 Shopping Center Research

THE BUSINESS OF DEVELOPING SHOPPING CENTERS inherently has enormous financial risks which tend to be proportionate to the size of the development. Because of these risks, many shopping center developers, owners, and investors seek to quantify the appropriate size of a development and design the optimal combination of tenants for the center as accurately as possible to ensure that it successfully serves trade area consumers. Shifts in a trade area's demographic profile, emergence of new types of competition, and the constantly evolving nature of retailing make it necessary to evaluate regularly the design and tenanting of a shopping center so that it continues to address its consumers' changing needs. This chapter addresses some of the more fundamental approaches to shopping center research (i.e., consumer research and feasibility analysis) that help in minimizing risk and making informed decisions.

Consumer Research

The purpose of consumer research is to obtain information that is useful in determining the feasibility, optimal tenant mix, and appropriate market positioning strategy of a shopping center by speaking directly to existing customers or trade area consumers. The insight that can be gained from consumer research is exhaustive and ranges from determining of the trade area extent of an existing center to identifying the tenant types that should be added or removed from a center. Several types of consumer research studies are commonly used in conducting shopping center research. The specific type used is a function of the nature of the shopping center (existing versus proposed) and the goal of the research (feasibility analysis versus a market positioning study).

Perhaps the most basic consumer research methodology is the customer source survey, which quantifies the geographic draw of a shopping center so that a trade area can be defined. The results of this research lays the foundation for the analyst to evaluate additional important issues such as trade area analysis or competitive impacts. Typically, a customer source survey for a shopping center consists of interviewing a representative sample of customers. The sampling would be conducted over a multiday period and the exact days and times of interviews would correspond to the level of business activity of the center. Interviewers at the center should be strategically positioned to intercept shoppers coming from the center's major anchors and its common area access points; oversampling in one area could distort the survey results as the customer traffic could be unduly influenced by the draw or appeal of tenants concentrated in these areas. Additionally, the interviewing process should be monitored to ensure that a random sample of the center's shoppers are surveyed.

For the purpose of a customer source survey, the interviewer only asks the respondents how much they spent cumulatively at all tenants in the center, and the place from which their shopping trip originated. The origin of the customer's trip may simply be their residence (as is typically the case with a suburban, supermarket-anchored neighborhood center). However, if the center is located near a tourist destination, a major office park, a Central Business District (CBD), or other nonresidential source of customers, it may be appropriate to also collect business addresses, seasonal residences, and so forth, as the point of shopping trip origin. As an example, if a customer source survey was conducted at the Crossroads of Lake Buena Vista Shopping Center just outside of

Walt Disney World, and responses were restricted to permanent addresses, the geographical distribution of customers would be international in scope. Very impressive, but useless in defining the trade area for this center. Many, if not most, of the shoppers patronizing this center do so because it is near their hotel.

The exactness of the information collected when conducting customer source surveys is a function of the nature of the shopping center. For regional malls or power centers which have very large trade areas, simply collecting a customer's ZIP Code of each customer may be sufficient. In the case of a neighborhood shopping center, however, the street address of each customer interviewed (or the street and closest cross street of each customer's residence) should be collected to achieve an adequate understanding of customer distribution since these centers typically serve small trade areas. The results of customer source surveys are typically plotted on a map to determine the center's trade area extent. The amount of the respondent's purchase is "tagged" to the respondent's trip origin so that the survey can be used to estimate the distribution of the center's sales, as well as its customer distribution.

A natural extension of a simple customer source survey is a detailed customer exit survey (see Chapter 12, Figure 12-1). In addition to the "trip origin" data that is collected during simple customer source surveys, the issues addressed by customer intercept surveys are as varied as the problems and opportunities encountered daily by shopping center developers and owners. Examples include customer shopping habits at the center such as the stores they shop, how often they patronize the center, the names of competitive centers they shop, and so on. In addition to behavior, customers can be asked questions regarding their perceptions (e.g., what they do and do not like) and what tenant additions or other improvements they would like to see at the subject center. Another use for customer intercept surveys can be to identify customer demographics.

Results of detailed customer intercept surveys are typically used to ensure that the subject center is addressing the needs of its customers, and can provide invaluable insight in formulating strategies related to retenanting, marketing, remodels, operational changes, and the like. As an example, the results can identify perceived the strengths and weaknesses of the center. Further, they will identify "holes" in the center's current tenant mix by providing an indication of where respondents are now shopping for goods and services not available at the subject center.

Consumer research information also allows the analyst to compare the demographic characteristics of the respondents with the tenant roster of the center to see if a good retail match is evident. As an example, many first and second tier suburban shopping centers have experienced a dramatic change in the demographic makeup of their shoppers over the years, but may not have recognized it. When these centers opened, they served a customer base comprised primarily of young families, but they may now serve an area dominated by empty-nesters with higher per-capita incomes. The implication of these findings on the center's tenanting, as well as the tenant's sales levels, are enormous. Without the insight brought into focus by the detailed customer intercept survey, the subject center could be in jeopardy of losing market share to a more appropriately positioned competitive center.

The mechanics used in interviewing respondents for customer intercept surveys are identical to those used for customer source surveys. However, while customer source surveys generally take less than a minute to complete, a detailed customer intercept survey is completed in about five or six minutes, although such surveys can occasionally last for up to twenty minutes.

Another valuable consumer research tool for an existing or proposed shopping center is a telephone survey (see Chapter 12, Figure 12-2). The most important distinction between a telephone survey and the intercept survey is that the telephone survey is typically used to elicit the opinions of randomly selected trade area residents who may or may not regularly shop the center, whereas intercept surveys can only address the opinions and perceptions of shoppers. As such, a telephone survey is useful in determining the attitudes and opinions of all consumers residing throughout the trade area of a proposed shopping center.

When surveying potential shoppers of a proposed center, the research analyst can gain information on a variety of issues, such as how far they travel to purchase the goods expected to be offered at the new center. Also, the research can qualify the strength of competing centers and the bearing they may have on the subject center. As an example, the research could determine that a competing center may have a very strong price image based on its anchor tenant roster. If the proposed center is much more upscale in nature, the analyst would be wise to temper the sales projection from the less affluent parts of the trade area near this competitive center. Finally, telephone surveys can be used to generate insight as to how potential customers will accept the new center. These insights will be determined by the respondents' comments on the new center's theme (off-price or upscale for example), projected tenant roster, and the like. If the plan for the center, and the respondents' opinions of this plan, are not in sync, consideration should be given to changing the scope of the new center.

Focus groups can be used to probe further concerns voiced by consumers during intercept or telephone surveys for either existing or proposed shopping centers. Typically, focus group research for shopping centers concentrates on issues that are not necessary to quantify. These issues might include perceptions regarding the center's ambiance (service levels, security, cleanliness) as opposed to where to locate a center or expected sales performance of proposed tenant types. When probing for tenant inclusion considerations in focus groups, the research analyst should be wary in that the opinions presented by the group are the opinions of a small sample of people (see Chapter 12, Figure 12-4).

Feasibility Analysis/Tenant Mix Analysis

The most comprehensive way to assess the feasibility of a new shopping center, or the retenanting of an existing center, is to conduct a sales forecast for every proposed tenant that may be in the center's tenant roster. However, this approach is not practical as it would be excessively expensive to develop a database for each prospective tenant. Also, the interested parties (i.e., the developers, investors, management, or advertising firms) rarely have access to the client-specific data necessary to conduct these analyses. Even considering a small neighborhood center with only ten tenants, the cost for developing the databases and conducting the forecasts could be hundreds of thousands of dollars, to say nothing about the time involved in conducting such an analysis. Therefore, time- and cost-efficient market research methodologies that are generic in nature have been developed for shopping centers. The general approach to establishing the economic feasibility of a shopping center involves establishing the retail expenditure potential within the trade area for each of the proposed tenant types, estimating the share of that potential the proposed tenant will be able to achieve, and calculating to corresponding sales volume that the tenant will achieve as implied by the estimated market share. At the heart of this methodology is the assumption that the shopping center is reliant on tenant synergy—that is, the drawing power and consumer appeal of the center's tenants will be complementary.

The first step in conducting a feasibility analysis for a proposed shopping center, or a tenant mix analysis for an existing center, is to thoroughly understand the dynamics of the trade area that the center is serving or will serve. Thus, before any analysis can be conducted, a trade area must be defined. As described above, a trade area definition for an existing center can be readily and accurately determined from a customer source survey. In the case of a new center, the analyst will need to define a trade area based on accepted location research theory. The factors typically examined when defining a trade area include:

- the deployment and strength of competitive centers, particularly "clone" centers;
- accessibility patterns to the subject site;
- population distribution;
- demographic distributions;
- physical or man-made barriers.

Once the primary trade area has been defined, the analyst can then conduct a field assessment of the trade area, followed by an analysis of the field data as well as data from secondary sources.

An in-depth field assessment is the first step in conducting a proposed shopping center feasibility study or a retenanting analysis of an existing center. Much like the fieldwork conducted for an individual retailer, the research analyst will be gathering primary information on the subject site, its trade area characteristics, and its competition. During the course of the fieldwork, the analyst should also meet with local sources to gather information on proposed competitive developments, road/access changes, and population growth. A major change to any of these criteria can have a substantial impact on the ultimate performance of the subject site.

With regard to site characteristics, the analyst will be critically assessing the site visibility, ingress/egress, traffic flow, and location relative to competition. In particular, the analyst will be closely assessing the traffic flow around the site and its location as it relates to similar competitive centers. Easy and safe vehicular traffic patterns around a site are tremendously important to the ultimate performance of the stores in the center. The analyst must answer a variety of questions, such as: Are the surface roads ample to engage and disperse traffic in all directions? Depending on the size of the center, is there adequate, close-by, regional accessibility so that the center can serve a sufficiently large geographical area? Will the center have enough convenient parking to satisfy the volume it is projected to achieve? Is the site proximate to an adequate number of in-profile shoppers? These points must be honestly evaluated in the context of the center's competitors if the analyst is to later produce a meaningful sales forecast.

Perhaps, the most important trade area characteristic evaluated in the field is accessibility. Specifically, the analyst will determine the proximity and condition of both local and regional access routes. Further, a check will be made to determine if there are any barriers, either physical or psychological, that will inhibit any parts of the trade area population base from easily traveling to the subject site. Besides accessibility, the analyst will also be making a qualitative assessment of the distribution of both population density and demographic characteristics within the trade area. These qualitative opinions will be later quantified when the analyst assesses secondary demographic data sources. Finally, the analyst will be assessing the deployment of existing and proposed shopping centers and major retail concentrations to determine how they will likely impact the trade area definition for the subject center.

In order to estimate the subject center's market share by tenant type, the analyst must have a thorough knowledge of the competition serving the trade area in each retail category. As such, the field analysis will include a thorough inventory of retailers operating within and immediately surrounding the center's trade area. This inventory is usually restricted to the retail categories that may be included in the center's tenant mix. Depending on the size and variety of tenants proposed for the subject center, this can be a very tedious task. As an example, if center's developer is considering 20 different retail types for the tenant roster (e.g., card/gifts, men's apparel, drug/HBA, etc.), the analyst will need to inventory each competitor to all of the twenty proposed tenants, at a minimum. In the interest of time and cost efficiency, the analyst typically counts the number of competitors in each retail category and gathers minimal additional information regarding their size or strength. As a part of the inventory, the analyst will have located all competitive shopping centers, as well as major, freestanding retail stores. Further, the analyst will have spent enough time at the competitive centers so that they can make knowledge-based comments on the tenant mix and philosophy of each center, as well as its size, strength, and merchandising philosophy of its anchor tenants.

Subsequent to the completion of fieldwork, the in-office analysis for proposed/existing shopping centers follows a very different course than that which was explained for the individual retail store. The in-office analysis begins with the determination of expenditure potential available for various retail types, and ends with the projected market share and corresponding sales volumes that the retailers in the subject center could be expected to achieve. As opposed to examining an extensive database and then calculating a forecast based on a well-defined application, a much more general and generic approach is usually taken. This is so not only because the analyst will have no client-specific databases to work with (and even if these were, the databases would be proprietary to the retailer—not to the developer or investor of the shopping center), but also because the identity of the retailers who will ultimately occupy space in the center may not be known at the time of the feasibility study. As an example, a developer/investor may want a bookstore in their center, but at the time of the feasibility study, they may not know if the ultimate tenant will be Borders Books & Music or Barnes & Noble, or if the analysis will determine that only a much smaller bookstore tenant will be feasible. Ultimately, the analyst will recommend a tenant mix and market positioning strategy for the center designed to maximize sales, taking into account the competitive environment and sales potential of the trade area.

In Chapter 13, a discussion was presented on estimating the expenditure potential for a start-up retail store. This methodology is replicated in shopping center feasibility and tenant mix analyses, only on a much larger scale. The expenditure potential for a retail type is the dollar amount available to be spent in the retail category by trade area consumers. As opposed to developing the expenditure potential for one retail type, the expenditure potential for up to several dozen retail types must now be developed. The expenditure potential estimates are based on information provided by government agencies such as the U.S. Census of Retail Trade. Other data sources for expenditure potential are state sales tax information and modeled expenditure data available from data vendors. Table 14-1 shows the type of retail expenditure data available from the U.S. Census of Retail Trade.

Whatever data source is used, the information is projected forward to account for any ensuing inflation between the base year of the data gathering, and the year for which the forecast will be conducted. Also, the population base of the trade area must be projected forward to the forecast year so that when it is multiplied by the expenditure potential, an accurate as possible estimate of total trade area expenditure potential is made. This procedure is calculated for each retail type proposed for the subject center.

After establishing the trade area expenditure potential for each retail type, the next step is to conduct a market share analysis. One common methodology used in this analysis is the share of space/share of market approach (explained in Chapter 13). This methodology is based on the theory that the proportion of space to be occupied by the proposed retailer (in relation to the total competitive space currently available in the trade area) will be equal to its trade area market share. The reader will recall that the analyst has already conducted an inventory of all the retailers in the trade area. Using this generic approach, the number of retailers inventoried in each retail category is multiplied by the average size of that retail type in order to approximate the total existing square footage devoted to that retail type. The average square footage factors are generally taken from a secondary data source such as ICSC publications or information from the Urban Land Institute. In some instances it may be appropriate to adjust the effective size of certain competitors in order to account for their individual strengths or weaknesses, or the strength/weakness of the shopping center in which they are located. For large-format retailers exact prototype sizes may be available from other secondary sources or from the retailer itself.

As an example in calculating market share, assume that a proposed retail tenant is a 30,000-gross-square-foot sporting goods store. Also, assume that there are 120,000-gross-square-feet of other sporting goods stores serving the trade area at the present time. Therefore, the sporting goods store at the subject site would have an assumed trade area market share of 20 percent (30,000 sq. ft./150,000 sq. ft.) assuming no increases or decreases were made to the strength of the retailer or the subject center. If the expenditure analysis for sporting goods had determined that, on average, each of the 325,000 people in this trade area spent $61 annually for these items (in the first year of the forecast period), then there would be a total expenditure potential for sporting goods of $19,825,000. Thus, the subject sporting goods store would capture 20 percent of that potential, or $3,965,000. If the store was further expected to generate 20 percent of its sales from beyond the trade area, then its total sales forecast would be $4,956,250 ($3,965,000/0.8).

In the above example, the forecast was conducted using a macro-approach; that is, the market share was calculated for the trade area as a whole. A more exacting way to conduct the sales forecast would be to subdivide the trade area into small pieces of geography, and

table 14—1

Table 1. Summary Statistics for the State: 1992

[Includes only establishments with payroll. For meaning of abbreviations and symbols, see introductory text. For explanation of terms and comparability of 1987 and 1992 censuses, see appendix A]

SIC code	Kind of Business	Establishments (number)	Sales ($1,000)	Annual payroll ($1,000)	First-quarter payroll ($1,000)	Paid employees for pay period including March 12 (number)
	Retail trade	**71 652**	**67 787 842**	**10 042 888**	**2 346 082**	**861 565**
52	**Building Materials and garden supplies stores**	**3 333**	**4 188 235**	**517 611**	**116 088**	**28 925**
521, 3	Building material and supply stores	1 675	3 128 010	369 462	84 404	18 809
521	Lumber and other building materials dealers	1 213	2 875 402	336 704	76 444	16 846
523	Paint an, glass, and wallpaper stores	462	252 608	32 758	7 960	1 963
525	Hardware stores	912	499 922	73 350	16 845	5 612
526	Retail nurseries, lawn and garden supply stores	577	347 952	52 391	9 878	3 389
527	Manufactured (mobile) home dealers	169	212 351	22 408	4 961	1 115
53	**General merchandise stores**	**1 631**	**10 166 652**	**1 134 097**	**262 979**	**101 242**
531	Department stores (incl. leased depts.)[1 2]	566	8 546 414	(NA)	(NA)	(NA)
531	Department stores (excl. leased depts.)[1]	566	8 289 766	969 503	224 869	85 592
531 pt.	Conventional[1]	123	(D)	(D)	(D)	KK
531 pt.	Discount or mass merchandising[1]	347	3 917 228	373 515	84 027	38 263
531 pt.	National chain[1]	96	(D)	(D)	(D)	JJ
533	Variety stores	667	545 495	69 899	15 947	7 476
539	Miscellaneous general merchandise stores	398	1 331 391	94 695	22 163	8 174
54	**Food stores**	**8 368**	**17 500 521**	**1 691 188**	**407 853**	**153 974**
541	Grocery stores	5 632	16 497 937	1 536 787	371 902	136 311
541 pt.	Supermarkets and other general-line grocery stores	3 167	14 907 804	1 377 651	335 575	117 284
541 pt.	Convenience food stores	1 563	958 188	106 114	23 638	12 566
541 pt.	Convenience food/gasoline stores	548	522 72	36 005	8 603	4 368
541 pt.	Delicatessens	354	109 218	17 017	4 086	2 093
542	Meat and fish (seafood) markets	597	378 085	38 945	9 178	3 362
546	Retail bakeries	1 000	265 449	69 780	16 585	8 840
546 pt.	Retail bakeries – baking and selling	886	233 810	65 081	14 458	8 154
546 pt.	Retail bakeries – selling only	114	31 639	4 699	1 127	686
543, 4, 5, 9	Other food stores	1 139	359 050	45 676	10 188	5 461
543	Fruit and vegetable markets	215	114 480	12 138	2 467	1 176
544	Candy, nut, and confectionery stores	346	69 011	11 479	2 830	1 874
545	Dairy products stores	164	38 863	4 835	1 074	605
549	Miscellaneous food stores	414	136 696	17 224	3 817	1 806
55 ex. 554	**Automotive dealers**	**4 339**	**17 840 677**	**1 424 607**	**322 321**	**60 245**
551	New and used cars	1 566	15 587 492	1 170 399	265 355	45 598
552	Used car dealers	868	768 132	50 841	11 499	2 967
553	Auto and home supply stores	1 493	1 005 860	157 032	36 067	9 299
553 pt.	Auto parts, tires, and accessories stores	1 443	985 228	154 468	35 504	9 110
553 pt.	Home and auto supply stores	50	20 632	2 564	563	189
55, 6, 7, 9	Miscellaneous automotive dealers	412	479 193	46 335	9 400	2 381
555	Boat dealers	104	94 350	10 520	2 076	524
556	Recreational vehicle dealers	122	197 409	17 093	3 592	790
557	Motorcycle dealers	167	176 283	17 212	3 371	995
559	Automotive dealers, n.e.c.	19	11 151	1 510	361	72
554	**Gasoline service stations**	**4 744**	**5 568 222**	**323 654**	**78 825**	**30 807**
554 pt.	Gasoline/convenience food stores	973	1 505 347	75 199	17 636	8 225
554 pt.	Other gasoline service stations and truck stops	3 771	4 062 875	248 455	61 189	22 582
56	**Apparel and accessory stores**	**6 732**	**4 540 846**	**511 440**	**119 983**	**51 616**
561	Men's and boys' clothing and accessory stores	739	488 792	70 994	16 457	5 044
562, 3	Women's clothing and specialty stores	2 755	1 624 714	188 406	45 073	22 005
562	Women's clothing stores	2 289	1 458 863	165 402	39 790	19 834
563	Women's accessory and specialty stores	466	165 851	23 004	5 283	2 171
565	Family clothing stores	718	1 206 491	108 481	24 846	11 303
566	Shoe stores	1 928	926 505	107 786	25 679	9 529
566 pt.	Men's shoe stores	173	64 603	8 983	2 226	619
566 pt.	Women's shoe stores	352	139 591	17 634	4 211	1 629
566 pt.	Children's and juveniles' shoe stores	78	20 664	3 476	834	327
566 pt.	Family shoe stores	1 081	514 843	59 186	14 008	5 313
566 pt.	Athletic footwear stores	244	186 804	18 507	4 400	1 641
564, 9	Other apparel and accessory stores	592	294 344	35 773	7 928	3 735
564	Children's and infants' wear stores	271	170 080	17 896	4 362	2 273
569	Miscellaneous apparel and accessory stores	321	124 264	17 877	3 566	1 462
57	**Furniture and homefurnishings stores**	**4 773**	**3 754 946**	**465 000**	**110 463**	**28 859**
5712	Furniture stores	1 413	1 253 122	173 188	41 359	9 803
5713, 4, 9	Homefurnishings stores	1 374	789 658	114 168	26 296	7 430
5713	Floor covering stores	649	460 529	69 235	15 863	3 536
5714	Drapery, curtain, and upholstery stores	113	32 477	4 461	1 055	435
5719	Miscellaneous homefurnishings stores	612	296 652	40 472	9 378	3 459
572	Household appliance stores	523	397 196	47 563	11 637	2 976
573	Radio, television, computer, and music stores	1 463	1 314 970	130 081	31 171	8 650
5731	Radio, television, and electronics stores	750	752 492	75 164	18 075	4 596
5734	Computer and software stores	183	219 790	18 448	4 176	897
5735	Record and prerecorded tape stores	352	243 621	20 758	5 222	2 256
5736	Musical instrument stores	178	99 067	15 711	3 698	901

See footnotes at end of table.

PA-8 PENNSYLVANIA RETAIL TRADE—GEOGRAPHIC AREA SERIES

Table 1. **Summary Statistics for the State: 1992–Con.**

[Includes only establishments with payroll. For meaning of abbreviations and symbols, see introductory text. For explanation of terms and comparability of 1987 and 1992 censuses, see appendix A]

SIC code	Kind of Business	Establishments (number)	Sales ($1,000)	Annual payroll ($1,000)	First-quarter payroll ($1,000)	Paid employees for pay period including March 12 (number)
58	**Eating and drinking places**	**21 063**	**8 177 872**	**2 123 609**	**495 823**	**282 212**
5812	Eating places	16 936	7 499 129	2 006 230	467 832	266 528
5812 pt.	Restaurants	7 887	3 761 367	1 092 390	256 252	143 464
5812 pt.	Cafeterias	137	61 425	17 281	3 924	1 948
5812 pt.	Refreshment places	6 872	2 757 792	645 394	148 841	93 387
5812 pt.	Other eating places	2 040	918 545	251 165	58 815	27 729
5813	Drinking places	4 127	678 743	117 379	27 991	15 684
591	**Drug and proprietary stores**	**2 813**	**4 266 821**	**447 258**	**104 881**	**31 598**
591 pt.	Drug stores	2 651	4 153 160	436 342	102 386	30 408
591 pt.	Proprietary stores	162	113 661	10 916	2 495	1 190
59 ex. 591	**Miscellaneous retail stores**	**13 856**	**11 783 050**	**1 404 424**	**326 866**	**92 087**
592	Liquor stores	1 452	1 374 457	98 090	24 009	5 085
593	Used merchandise stores	612	169 779	31 199	7 340	2 976
594	Miscellaneous shopping goods stores	5 793	2 838 822	358 342	83 852	34 945
5941	Sporting goods stores and bicycle shops	1 010	560 837	64 327	14 418	5 486
5941 pt.	General line sporting goods stores	359	288 262	31 206	7 019	2 750
5941 pt.	Specialty line sporting goods stores	651	272 575	33 121	7 399	2 736
5942	Book stores	569	318 674	37 025	8 733	3 857
5944	Jewelry stores	1 269	582 682	91 480	22 142	6 796
5943, 5, 6, 7, 8, 9	Other miscellaneous shopping goods stores	2 945	1 376 629	165 510	38 559	18 806
5943	Stationery stores	171	61 600	10 165	2 583	921
5945	Hobby, toy, and game shops	537	533 533	47 122	10 845	5 044
5946	Camera and photographic supply stores	177	96 201	13 578	3 289	972
5947	Gift, novelty, and souvenir shops	1 631	494 206	67 630	15 497	8 732
5948	Luggage and leather goods stores	100	43 321	6 456	1 502	539
5949	Sewing, needlework, and piece goods stores	329	147 768	20 559	4 843	2 598
596	Nonstore retailers	1 180	4 759 839	510 985	114 876	23 233
5961	Catalog and mail-order houses	289	3 971 887	345 424	74 285	12 561
5962?	Automatic merchandising machine operators	279	398 217	87 356	21 990	5 091
5963?	Direct selling establishments	612	389 735	78 205	18 601	5 581
598	Fuel dealers	756	1 368 895	164 321	40 521	7 090
5983	Fuel oil dealers	587	1 213 059	142 219	35 365	6 163
5984	Liquefied petroleum gas (bottled gas) dealers	132	145 641	21 238	4 945	852
5989	Fuel dealers, n.e.c.	37	10 195	864	211	75
5992	Florists	1 340	301 049	65 829	15 932	6 477
5993	Tobacco stores and stands	92	35 460	3 200	717	337
5994	News dealers and newsstands	252	64 805	6 888	1 647	842
5995	Optical goods stores	702	242 763	54 538	13 456	3 231
5999	Miscellaneous retail stores, n.e.c.	1 677	627 181	111 032	24 516	7 871
5999 pt.	Pet shops	331	115 505	16 628	3 979	1 941
5999 pt.	Art dealers	143	48 342	9 194	2 094	563
5999 pt.	Other miscellaneous retail stores, n.e.c.	1 203	463 334	85 210	18 443	5 367

[1]Includes sales from catalog order desks.
[2]Includes data for leased departments operated within department stores. Data for this line not included in broader kind-of-business totals.

then estimate the market share for each of the subsectors in light of the competitive environment serving each subsector. This methodology will allow the analyst to better account for the distribution of competition, as well as demographic characteristics. Further, the analyst will also be able to account for distance decay in market share.

The final step in the feasibility analysis would be to calculate the sales per square foot implied for all the forecasted retail stores. Continuing the above example, the proposed sporting goods store is expected to achieve $165.21 in sales per gross square foot ($4,956,250/30,000 sq. ft.). This figure should be compared to industry data or actual store operating data (if available) to determine whether the sales forecasted are sufficient to support the store. If the sales figure is robust enough, the retail type should be recommended for inclusion in the subject center's tenant mix. If not, the retail type should probably be rejected. Once this analysis has been conducted for all retail types, the analyst can then determine which of the retail types appear to offer the most potential for the subject center. This analysis can also aid the analyst in quantifying the most appropriate size for a shopping center, as well as the tenants it should include.

As one can imagine, the calculation of market share for several dozen retail types over several pieces of subgeography can be quite painstaking. Because of this, and because of the associated time and cost considerations, it may be more expedient to conduct anchor tenant-based retail feasibility analysis as opposed to an all inclusive tenant feasibility analysis. The main types of anchor tenant-based feasibility analyses are defined below:

- **Regional Mall**—A sales forecast is conducted for the combined space devoted to department stores at the subject center incorporating the methodology described above. The field evaluation focuses only on department stores, discount department stores, and junior department stores that compete for available expenditures in the general merchandise retail category within and immediately beyond the trade area of the subject center. This analysis is also useful in determining the potential to support an additional department store anchor as part of a mall expansion.

- **Community Center**—A forecast is conducted using the methodology described above for selected generic retail categories (a large-format sporting goods store, bookstore, toy store, home center) rather than for specific operators. Conclusions regarding the feasibility of the center as a whole are tied to the projected performance of the anchors.

- **Supermarket-Anchored Center**—A sales forecast is conducted for a supermarket anchor using the gravity model forecasting methodology which considers the available expenditure potential and supermarket competition impacting the site. Conclusions regarding the feasibility of the neighborhood center as a whole would be a function of the feasibility of the supermarket anchor.

Many of the processes used for individual store forecasting are used in shopping center research, and the analyst will still be critiquing site characteristics, analyzing competitors, and studying demographic/population data. Further, the whole process is, ideally, augmented by consumer research. The major difference between store-specific research and shopping center research is the more general (or generic) nature of the sales forecast. Nonetheless, because most of the supporting data was gathered using sound location research principles, the sales forecast for the new or retenanted center, and the implications derived from the forecast, will be both meaningful and supportable.

15 The Future of Site Location Research

FOR SEVERAL DECADES THE PIONEERING WORK OF Applebaum and Huff has provided the basis for store location theory and practice. As in all fields of research, continual technological and methodological advances have fostered the evolution of the site location and evaluation process in the retail industry. In the last several years, this evolution has been taking place at an increased pace. This rapid advancement has occurred largely as a result of increases in the power and availability of computers and their associated software. This final chapter presents some of the evolving ideas and processes that are now being used and evaluated in store location research.

Three trends will shape store location research in the next few decades. Perhaps more importantly, however, is the synergy created by their interaction. The first important location research trend is the significant increase in the efficiency of collection of data and the development of databases. For many years, advanced spatial interaction models have been widely used in the academic world, but have not been extensively used in store location research. Why have simpler methods prevailed when more sophisticated techniques have been available? The answer lies in a cost-benefit analysis of using such techniques—the data necessary for such techniques has been unavailable or prohibitively costly to develop. Academic use has been possible only by grossly simplifying data gathering. This approach is an unrealistic and unacceptable compromise in store location research where significant investments are at stake. Thus, the pragmatic consideration of the costs of developing the database needed to apply advanced techniques has inhibited their use. As the process of developing massive databases becomes more efficient, particularly if advanced data structures discussed later in this chapter such as data warehouses are employed, it is likely that more sophisticated techniques will be commonly used.

The use of advanced statistical and modeling techniques, such as spatial interaction models, represent the second area where rapid advancements are taking place. A richer data environment enables both more sophisticated statistical analysis and the development of increasingly more complex computer-based systems to support locational decision-making. There is a close interaction between the ability to develop an appropriate database and the use of more sophisticated analytical techniques.

The third major trend will be the continued rapid evolution of geographic evolution of geographic information systems (GIS); an enabling technology that fosters spatial visualization and more efficiently integrates data necessary for location research. GIS provides the location analyst with the ability to integrate the databases, analytical tools, and to visualize the resultant spatial patterns.

The importance of these three trends shaping location research is their synergy, truly a case where the result is more than the sum of its parts. The glue that holds these interdependent trends together is the availability of powerful computer systems and programs. In a similar vein, strategic analysis tools can now be used in a much wider variety of contexts. The more widespread use of computers has in turn resulted in a much wider market for data products. Finally, availability of data and increases in computer processing speed and storage has enabled the rapid growth of GIS use by businesses. Computer processes that would have required super-computers several years ago can now be accomplished on the desktop. This widespread availability of computer power has also increased the speed with which advanced statistics and computing algorithms have emerged from the academic environment to the practical world of store location research.

ADVANCES IN DATABASE DEVELOPMENT

Customer Information

Central to the development of most retail sales forecasting databases is the process of defining trade areas for stores which comprise the database. Among factors influencing trade-area size and configuration are access, demographics, competition, and site characteristics. Determining the trade area results from an analysis of each database store's customer distribution, either by gathering customer information directly or indirectly. Historically, collecting customer distribution information has been a very expensive process. A typical customer intercept survey can easily cost $1,200 to $1,500 per database store to gather 300 valid interviews. Extended out to a sufficient sample of stores, just generating this one component of the database can become prohibitively expensive.

Fortunately, many retailers have been able to leverage their other investments in data processing to greatly facilitate the collection of customer distribution information. The first of these methodologies harnesses the point-of-sale cash registers to capture customer information during the checkout process. This procedure was adopted several years ago by a few of our large format retail clients. Since their trade areas were large, ZIP Code geography provided a suitable level of geographic disaggregation for the purpose of trade area definition. By implementing some relatively minor reprogramming of the POS software, these retailers were able to simply capture the "home" or workplace ZIP Code of a representative sample of customers.

Another means of capturing information is the use of "Customer Appreciation" cards, which provide merchandise and coupon inducements in exchange for personal information. For example, many retail companies have now developed or are developing customer data warehouse structures (i.e., exceptionally large customer databases) to facilitate their "one-to-one" marketing efforts. The phrase "one-to-one" generally indicates an effort by the retailer or service provider to market directly to individual customers based on that customer's prior purchasing habits or the purchasing patterns of customers with similar profiles. There is a growing trend for retailers to combine POS data capture with sophisticated data architecture to achieve maximum impact from customer information. If part of this effort included collecting exact customer addresses (as opposed to simply collecting ZIP Codes), these databases can provide a rich source of customer information for a variety of analytical purposes.

A similar trend in the supermarket industry is "preferred customer" membership programs. These programs typically aid in the check validation process, provide the customer with access to special promotions, and provide the retailer with very useful information about the customers. As retailers continue to collect more data about their customers and assembling the data into large databases continues, there will be increasing opportunities to use this data for locational research. Michael Winch, Information Technology Director for Safeway Stores plc (UK) recently reported significant gains in same store sales achieved through the full integration of a "preferred customer" program throughout their chain (Keynote Address, Retail Systems 97 Conference).

In addition to internal sources of customer information, external sources of customer information, such as credit card and credit bureau information are also becoming available. For a fee, some credit card companies will now provide retailers with information such as geocodes, and limited demographic indices on their customers. Like the internal databases described above, this provides retailers with opportunities for marketing directly to individual customers based on their prior purchase patterns, as well as providing the customer distribution necessary for location research. Credit reporting agencies can provide similar fee-based services to identify the distribution, demographic characteristics, and major purchase decisions of retail customers. However, some cautions should apply to this information. First, these are controversial business practices, particularly with regard to privacy issues. Thus, access and use of such data may be restricted as car registration data is now in many states. Secondly, also as with car registrations, the geocoded information may not be current, due to the inherent lags in the reporting processes. These potential data problems have the effect of adding to the overall error rates in statistically based forecasting processes.

The data-processing trend that has made much of the above customer information possible is geocoding software. The large-scale emerging of geocoding as a factor in customer and consumer research is a result of a combination of more powerful data resources, the increased ability of microprocessors to manipulate large data files, and the software to control the process. National geocoding files first became available as a result of the Census Bureau to streamline their data collection processes. What eventually resulted was TIGER (Topographically Integrated Geographic Encoding and Referencing) which provided two important benefits to location research; comprehensive national files for both address ranges and cartographic line files representing roads, boundaries, and hydrology features. This enabled geocoding, the process of identifying a location as a set of coordinates based on another means of location description, such as an address or cross streets. Using geocoded files allows businesses to add a spatial component to their existing data files and further to relate other existing data files containing spatial data, such as demographic characteristics or competitor locations. This can greatly streamline the process of developing customer databases. Over the past few years, the geocoding process has been further refined by the development of address standardization software that significantly increases the ability to precisely locate customers. Other trends in the geocoding industry have included unbundling the geocoding information from the cartographic data that result in a more simplified and efficient matching process, and the rise of companies devoted to cleaning and refining the quality of geocoding data. The most recent trend is embedding the geocoding process into other systems. It has become more common to see geocoding incorporated into on line analytical processing (OLAP) tools to build customer databases.

Other related trends are also contributing to the capacity to capture data in more efficient and cost effective manners. Global positioning systems (GPS), first developed in military contexts and now integral to all types of mapping, are finding their way into the field of location research. Rather than a process of gathering data in the field, transcribing it onto coding forms, geocoding the locational coordinates, and then starting the analysis, connecting the GPS directly to a laptop allows the instant digitizing of competitor locations. Further, connecting this system to the corporate computer network via cellular phone and modem will allow the instantaneous updating of competitor databases. Other innovative uses of technology also may increase the efficiency of collecting raw data. Touch screen monitors offer a method of gathering self-administered survey data. These systems are already used for gathering customer satisfaction data and may be adaptable to other types of surveys, if sample reliability issues can be successfully overcome.

Quality of Data Sources

Marked increases in the quality of commercially available data have occurred concurrent with the advances in efficient data gathering discussed above. As with many of the other innovations discussed in this chapter, this has been the direct result of the leverage of computer power. Particularly, the quality of competitor databases has seen a marked increase in recent years. While such data has been available for many years, recently, the data vendors have developed partnerships with their own customers, as well as their traditional supplier and grocery wholesaler sources. These partnerships create feedback loops to continually refine and update their sources of raw competitor data. In addition, demographic data vendors, companies who repackage and augment census demographic data, have branched out to market a variety of specialized spatial databases such as crime, health care, and automotive registration data. Finally, with budgetary constraints limiting the quantity and variety of data available through the Census Bureau, demographic vendors will step forward to fill the data voids albeit at a much higher price; the era of high-quality government-subsidized data has probably ended.

With the burgeoning growth of the world economy, sources for demographic and cartographic data have increased dramatically, giving rise to the diffusion on a new type of data resource, *metadata*. Metadata is data about data; its source, currency, quality, and suitability for a given purpose. In the past, such cataloging of data sources and availability has been a strictly do-it-yourself activity. With the rapid increase of the quantity of available data, keeping track of the latest in data developments has become a full-time job in itself. It is now fairly common for metadata generation and access to the resultant data library is a utility program commonly available from companies specializing in data warehousing software. The means of disseminating metadata ranges from traditional reference books to the most current Internet web sites.

The Internet is a particularly powerful means to extend the reach of locational research. It provides a means to go directly to the source of the data, in ways that both parallel traditional library resources and with powerful new search engines. One such example is finding or confirming competitor locations. It used to be that there were only two sure ways of confirming the locations of competitors, either by field research or by telephone detective work. Many retail and service firms now use the Internet as a means of providing information to their customers. This information has value to the location research process. For instance, a potential customer could visit a retailer's web site and make an inquiry as to the closest outlet. This same facility can be used for competitive research; all locations can by systematically downloaded and address geocoded to provide an up-to-date list of all locations. In a recent example from our business, a client wanted a national map depicting the locations of one specific competitor. One national list vendor could provide 700 locations and another 750. Using the Internet to visit the competitor's own web site, we discovered 825 current sites and 35 future sites complete with planned opening dates. In a similar vein, Internet research turned up sufficient demographic information for prospective sites in foreign countries and eliminated the need for an in-person visit to a specialized data library.

Data-related trends are not all positive, however. Recently, there has been a consolidation of demographic data vendors, which lessens the intensity of competition. The full ramifications of this consolidation have yet to be realized, but could result in less pressure to innovate and decreased service levels. Another threat may be the increase in the Internet's popularity. Already there has been significant degradation in access times to popular web site destinations. Without significant investment in increasing available bandwidth between server sites and in connectivity between individual users and Internet portals, the information superhighway will suffer data "traffic jams." Finally, the Bureau of the Census may institute significant cutbacks in data gathering and dissemination with the year 2000 census. While it is likely that data vendors will step forward to fill the voids, it is equally certain the end-user data costs will also rise.

Trade Area Definitions and Market Analysis

With greater access to customer data, particularly through the use of POS customer surveys and customer data warehouses, trade area analysis may take on an added dimension. These new methodologies allow an existing store's trade area to be defined based on an analysis of the distribution of its overall sales. Beyond this basic application, POS surveys also allow connecting customer "point of shopping trip origin" information with merchandise category information (in the form of SKU codes), thereby opening new dimensions to locational analysis. In particular, a retailer would be able to conduct a very detailed analysis of the relationships that exist between the purchase of specific merchandise and factors related to the purchase. For example, the analyst would be able to determine who buys specific items of merchandise, their location relative to the store, their demographic profile, when they buy, and what other types of merchandise are purchased concurrently. Based on this level of understanding, the effects of marketing activities can be analyzed at a very detailed level. For example, the impact of specific promotions, the placement of merchandise within the store, the interaction of purchase decisions and demographic characteristics of individual customers, and matching a store's merchandise mix with the demographic composition of its trade area are a few of the types of analysis that can be conducted. This type of detailed analysis is becoming fairly common in the supermarket industry. Increased availability of the data and the tools with which to analyze this information guarantees the widespread use of these types of analytical tools by other types of retailers in the future. This analysis will be extended to both new and existing stores. In recent years, it has not been uncommon for retailers to use a variety of merchandising prototypes within their stores. In the future, not only will each store's merchandising strategy be customized to its trade area, but the merchandising strategy will evolve as a result of the ongoing analysis of detailed customer behavior data. This detailed data will also impact sales forecasting. The distribution of trade area demographics combined with a detailed understanding of the interaction of merchandise assortments, demographics, and lifestyle will become components of the sales forecast. A natural extension of this analysis is the determination of what type of merchandise lines will sell in a particular trade area or even which departments should be represented in a store. Further, transfer impact on sister stores will help to fine-tune the merchandising presentation for an entire network of stores.

Accessibility Measures

The ability of a retail outlet to attract customers is at least partially a function of the accessibility of that store relative to alternative shopping opportunities. While this is most easily measured in terms of straight-line distance, in most instances this provides a very poor model of shopping behavior. Accessibility is more accurately measured in terms of driving distance or time. Unfortunately, sometimes the determination of which type of measure to use is more a result of cost considerations than what is the most appropriate measurement for the situation. As an example, because of the necessity of making measurements to a large number of sectors, straight-line distance is often used as a surrogate for accessibility in spatial interaction models. If most urban areas were laid out in precise grid road systems and there were no physical barriers, then straight-line distance would be a suitable measure of accessibility. Of course, this is never the case. Recent innovations in GIS technology allow more suitable measurements in an economical manner. In order to support more advanced time-distance calculations, however, digital road files including speed limits, traffic controls such as signals, stop signs or restrictions, directional indicators, number of lanes, and impedance factors (resistance to increased flows) are necessary. Thanks to investments made by companies such as ETAK to support in-car navigation and routing systems, this type of detailed road network data is now becoming available in a few test markets. As the tremendous investments necessary to create such information are made, the necessary data will be come widely available for location research, as well as other purposes. Of course, such data will remain at premium prices until its market expands sufficiently. The models and processing software to perform advanced accessibility calculations are currently available within a variety of GIS environments, but the data to support the calculations on a widespread basis, is not. It may be tempting to perform such calculations today, but without a thorough understanding and confidence in the accuracy of the underlying data, the results are likely to be disappointing.

Reliability Measures

With increased robustness of the data from which models can be derived and increased demand for automated forecasting systems, more sophisticated measures of forecast reliability will be demanded in the future. After all, if sales forecasting is viewed as a risk-abatement process, a more precise assessment of the reliability of the forecast process is entirely reasonable particularly in an automated forecasting environment. As customer samples available to the researcher approach 100 percent, and the connection between individual shopping behavior and the customer's demographic and lifestyle characteristics becomes more certain, the relationships that can be established become less inferential and more factual. This will enable more accurate forecasts and more reliable assessment of forecast risks. The measurement of forecast risk is determined by the underlying forecast methodology. For statistically based forecasting such as regression, risk assessment is an inherent part of the procedure and is measured as the error rate for either individual disaggregate predictions or for all predictions for a given store. In analog forecasts, risk can be assessed by relatively straightforward measurement of how closely the forecast situation matches the analogs used in its forecast. Typically, reliability is measured by the statistical similarity of the forecast to the analogs used to produce the forecast. In nonparametric statistical approaches such as CHAID (discussed further below), reliability is commonly measured as the percent of correct predictions.

Advanced Statistical Techniques

Several statistical techniques have promise for increased use in locational research. Three in particular, neural network/artificial intelligence tools, CHAID or Chi Square Automatic Interaction Detection, and spatial allocation models hold the greatest promise in the near term.

Neural networks are computer systems inspired by their biological counterpart, the brain. The specifics of how neural nets are developed and applied are quite complex. Full discussion of this technology is beyond the scope of this text. Suffice it to say that neural networks, when properly developed, are able to assess a database of stores, determine the interrelationship between sales and variables impacting sales, and can be "trained" to produce sales forecasts at new locations.

Neural nets may be appropriate for many classes of prediction problems where the relationship between the input variables and the values to be predicted is well understood. They are not, as some proponents have suggested, ideally suited to data exploration problems for two reasons. First, the internal processes in the neural net procedure are unknowable; it is a fitting process with no analytical output. Therefore, the logic of the prediction is unavailable for review. As a result, its suitability as a forecasting system is entirely dependent upon one's confidence in the underlying technology and the skills of the practitioner applying the technology. Secondly, the quality of the solution is a direct function of the quality and suitability of the input data. As with so many other things, the cliché "garbage in, garbage out" applies. Some proponents of the technique advocate loading in large amounts of data for analysis to determine if neural techniques can uncover "unknown" relationships. As with statistically based data exploration techniques, the quality of the thought in planning the research is usually directly related to the suitability and serviceability of the end product. Unless accompanied by a rigorous critical evaluation of neural net results, satisfactory forecasting results will not be realized.

Neural nets have been successfully applied to a variety of forecasting situations in retail such as inventory, purchasing, and promotions planning. However, the experience in sales forecasting for retail locations has been far less fruitful thus far, failing to equal, let alone significantly improve upon, other techniques that are simpler and far less costly to calibrate and implement. Further, neural networks require extremely large datasets of very clean data to solve complex problems. In summary, while there is promise in the technique, results in real-world forecasting situations have been thus far disappointing.

Many problems involving decision making, can be resolved using expert systems. A significant advantage of expert systems is their ability to provide detailed explanations of the process by which the solution was obtained. Expert systems utilize knowledge typically gathered from extensive structured interviews with "experts" in the application area with inference mechanisms based on formal logic, thereby enabling the program to utilize the knowledge for problem solving. In other words, expert systems seek to replicate the processes used by specialists; their thought processes, the data they use, how they resolve conflicting evidence provided by the data, or "rules-of-thumb" commonly employed by experts to approximate solutions. When successful, expert systems embody the experience and judgment of the specialist. Expert systems differ from conventional procedurally based programs in several important ways, most notably in that expert systems are not linearly processed; rules are not executed in any set sequence. Expert systems are most appropriate where knowledge is fragmented and implicit, and decision rules are complex and conditional within the context of the application being solved. Expert systems can also involve "fuzzy logic" processes, where conditional statements are resolved in probabilistic rather than absolute terms. In summary, expert systems are most appropriate where a combination of facts, judgment, and processes combine to form a solution.

In retail location research, expert systems are most often used in combination with other forecasting systems, particularly in automated systems as discussed later in this chapter. With greater understanding of the possibilities presented by combining emerging technologies, there will be increased demand for systems that capture forecasting expertise. Expert systems technology will be a necessary part of such systems.

CHAID or Chi Square Automatic Interaction Detection is a methodology that has been in existence for quite some time but has not been widely used due to the difficulty in calibrating the models. Recently, however, several statistical package vendors have assembled relatively easy-to-use CHAID calibration systems. CHAID has significant advantages for restaurants or convenience goods retailers where complex interactions between site and situational characteristics play a major role in determining a site's sales potential. CHAID provides a type of segmentation modeling that divides observations into segments that differ with respect to a designated criterion such as sales or market share. The observations are divided into two or more distinct groups based on categories of the predictor variable. In a location research context, the predicted variables might include total sales, per capita sales, or sales per square foot. Predictor variables might include income level, strength of competition, or population density, much the same as the variables that might be used in an analog or regression forecasting system. The division process continues until no further significant subgroups can be created. The final subgroup configuration is typically then displayed as an easy-to-understand tree diagram. The segments that are derived are mutually exclusive and exhaustive such that each observation belongs to only one subgroup. In application, each new observation can be readily classified into its appropriate segment just by knowing the category designations of the predictors. CHAID differs from traditional cluster analysis in that it utilizes a dependent variable as the criterion for defining the subgroups. CHAID is a nonparametric technique, meaning that it operates using categorical and dichotomous (yes-no) rather than interval (continuous) variables. The practical implication of this is that there is no limitation on the number of segments that can be created in comparison to forecasts based on parametric statistical techniques such as regression. Regression equations can support a very limited number of predictors for a given sample size. A more subtle result of not having these limitations in CHAID is the possibility of creating several independent decision tree structures out of the same dataset. This means separate decision tress that, for example, could focus on competition, the customer profile, and site characteristics could be created, resulting in the ability to take advantage of convergent forecasting processes.

In location research, CHAID analysis is particularly appropriate for use where complex interaction effects may be expected such as for restaurant, convenience, or mall-based retailers. For example, CHAID is capable of assessing the impact of a variety of adjacent cotenancies, or the interaction between specific target customer groups when certain competitors are present. If knowing which of two alternative locations within a mall provides the more suitable alternative is important, CHAID provides a powerful solution. CHAID is particularly suited to assessing complicated site characteristics or competition scenarios.

Most of the location research processes used today are not capable of simultaneously addressing more than one site at a time. While these techniques can be extended to incrementally consider multiple site scenarios, the work and calculations involved tend to increase more or less exponentially. From a strategic planning perspective, there is a growing need to simultaneously evaluate an entire network of sites in order to determine how sites may interact with one another, compete with the existing competition, and ultimately, determine an optimal network of facilities for a given market situation. Spatial allocation procedures are ideally suited to these types of situations. To date, however, spatial allocation has only rarely been used in applied retail location research due to its focus on examining the entire marketplace at once and the resultant high costs of data and calibration.

Fundamentally, spatial allocation procedures allocate spatially distributed demand to a specific number of retail outlets. Gravity models, discussed earlier in Chapter 9, are simple variants of spatial allocation models. While the primary focus of research in this area has been in the public sector, where adequately describing both the supply and demand functions tends to be simpler (i.e., public services such as fire stations do not have a lot of competition), the standard solutions can be readily modified to fit many varieties of optimal location problems. As was briefly touched upon in the introductory comments of this chapter, the practical problem of location allocation has always been providing input data suitable to the problem.

The first step in spatial allocation is to determine a suitable unit of geography within which demand can be distributed. The market would typically be subdivided into standard geographies such as tracts, block groups, or ZIP Code areas. Selecting the appropriate scale of geography is a critical decision because each submarket area is treated as a point within the model with the explicit assumption that the demand function can be realistically represented. Obviously, if the demand for the goods or services in question is demographically segmented, then the accuracy with which demand can be estimated becomes a function of the homogeneity of each subarea. The spatial allocation problem is typically configured to locate a specific number of outlets to best serve the overall market. The spatial allocation algorithm exhaustively evaluates every possible location within the market to determine the most efficient set of alternatives. Efficiency is usually measured in terms of minimizing the aggregate distance or cost of the total network. By making assumptions about consumer behavior, the program can use maximum consumption, sales, or profitability as the efficiency criterion. Modern spatial allocation algorithms offer flexibility in configuring the forecast scenario, allowing some locations to be fixed, considerable flexibility as to the specification of the demand functions, or the ability to more accurately represent the efficacy of competitors.

The problem with spatial allocation has never been the flexibility or sophistication of the spatial allocation software, but the difficulty of supporting the model with realistic data and a tendency to oversell its capabilities by overzealous practitioners. Spatial allocation requires both detailed inventories of the demand and supply relationships for an entire market area. On the supply side, the optimal index of competitor performance is each competitor's sales volume, which, for many types of retailers, is difficult to obtain. Firms with very specific and well-defined customer profiles may encounter significant difficulties in adequately specifying the demand relationships. If the retailer has multiple store prototypes or there are simply too few stores available for the modeling process, the accuracy of advanced modeling capabilities will be limited and probably not justify the high expense of calibration. Even larger chains with sufficient data will weigh the size, complexity, and expense of sophisticated models against their perception of what is cost effective for their particular chain and the challenges it faces.

Geographic Information Systems

In the last several years, Geographic Information Systems (GIS) have come of age. According to Geographic Technology Market (Fort Collins, CO, GIS World, 1995), business applications such as site selection and market analysis have become the leading growth market for GIS. The three critical areas in GIS systems are data, software, and hardware. In each of these areas, there have been significant changes that enable efficient utilization of GIS systems by retailers.

While GIS technology has been available for almost 30 years, its early use was confined largely to government, utilities, and natural resources industries where it was focused on managing "mission critical" resources. Data is the lifeblood of any information system. For these industries, the costs of primary data generation necessary to support GIS functionality was easily justified. Business applications such as location research, however, could not justify the high costs of primary data and therefore did not embrace GIS technology until low-cost data became widely available after the release of the 1990 census. More recently, low-cost comprehensive databases have been developed and expanded upon by a large and growing number of data suppliers. These suppliers offer their products configured in a variety of common GIS formats. This trend, which has achieved considerable momentum since 1990, is largely responsible for the growth of GIS usage in location research. Unlike the early adopters of GIS technology, location research users were not dependent on highly proprietary data with limited applicability to other users, but rather utilized data components such as demographics, road, and competitor location files that had applicability to a broad range of users. As GIS usage expands into a wider array of business, the market for data will continue to attract new providers and products, which should produce more competition, increased quality of data products, and lower costs. As an example, just a few years ago, a comprehensive national road file cost in excess of $100,000. Now, comparable files can be obtained for around $10,000 and files of somewhat lesser quality for less than $1,000.

As with private vendors, government data providers have also greatly increased the quantity and availability of data sources. United States government data has historically been very inexpensive, but lately there has been pressure to achieve better cost-recovery ratios from data end users. If this happens, and it seems likely that it will to some degree, government data costs will increase dramatically as has been the case in Canada where such programs are already in place. As government-supplied data becomes significantly more expensive and specific data items are no longer available, private data vendors can be expected to move to provide a larger share of data needs. On more local levels, there has been a very positive trend of governments becoming more willing to share data sources. However, this data sharing is not at all coordinated on regional or national levels resulting in great difficulty in determining data availability and quality levels.

Going forward, satellite and aerial imagery will play an increasing role in location research. This is due both to increased availability and, with more widespread use, lower cost. Further, the quality of such imagery will improve dramatically with the increased availability of declassified satellite images. Currently, aerial imagery is typically used in site analysis to convey a sense of locational context. It is also used in conjunction with high-end GIS mapping where the tools to integrate and rectify such images with vector map formats is quite advanced. Advanced image manipulation and integration tools are now becoming available on GIS platforms for the personal computer. These tools, in combination with submeter resolution imagery, will have great utility in location research. It will both largely replace conventional aerial photo imagery and can be readily used to verify development patterns and housing count data, the type of due diligence that now requires fieldwork.

Until very recently, utilization of GIS for anything beyond simple map-making required a technical specialist. Perhaps the most significant change in GIS software is the migration of high-end GIS functionality now provided by PC GIS vendors. In addition, there has been a move toward standardization in programming languages and toward "windows compliant" approaches to information interchange and away from proprietary standards. This downward migration has taken place with platforms also, both ESRI and Intergraph now offer their high-end GIS on the Windows NT operating system. Many retail chains in particular have been reluctant to embrace GIS on UNIX workstations due to the costs of supporting an additional operating system platform. Retail chains are now aggressively moving toward Windows NT to combine the familiarity and flexibility of the PC environment with the "industrial strength" processing power formerly only available on client-server or mainframe environments.

Major trends in the software industry such as object-oriented programming, client-server computer architectures, the Internet, and open systems have counterparts in GIS software. Location research is not a standardized discipline and, as a result, retailers have their own unique requirements as to reports and analytical processes required for site approval. As a result, a "shrink wrap" approaches to GIS-based locational analysis have not achieved widespread acceptance; highly customized approaches are usually required. At the same time, GIS application development has become increasingly complex and time consuming. Further, developers with experience in new extensions to GIS technology such as multimedia, imaging, and 3-D graphics are scarce and expensive human resources. Fortunately, object-oriented programming environments enable developers to readily reuse common application components. These components can encompass all aspects of the application, including the domain-specific functionality, business rules, report and map generation, and much more. PC-based GIS applications are now providing class library technology. This offers the developer the possibility for acquiring off-the-shelf technology that can be readily modified and effortlessly incorporated into GIS applications, thereby improving both the quality of the application and developer productivity. In summary, the continued adoption of object-oriented technology will bring a variety of benefits to GIS productivity including reusable components, availability of commercial "plug-in" class libraries, and compatibility with full-featured, object-oriented languages.

GIS has been adopted into retail organizations primarily through the real estate and logistics functions. These initial applications are now growing into other departments as shared data resources are becoming enterprise-wide applications. As the evolution from individual to departmental to enterprise applications has taken place, the development process itself becomes more complex. Large development projects built by teams of developers require version controls, the ability to maintain links between modules, and consistent current documentation. Without such code management, multideveloper projects cannot succeed. The GIS platforms of the future will provide means of managing the development process such as an application repository. A GIS program repository will allow better management of GIS projects through development and maintenance, provide management of metadata, provide browsers with a means to quickly locate the code within the repository, and provide check-in and check-out of program objects, as well as provide version control.

In the near term, these third- and fourth-generation language components will be used to prototype and develop GIS applications at a rapid pace. As these systems mature, large components that act as intelligent "agents" (i.e., computer programs that develop other computer programs) for developing GIS applications will guide the entire development process. The end result of such agents is that developers can focus more on application logic and allow the agents to integrate database, network, and middleware components of the application. At the same time, the corporate network will be extended to incorporate advanced network models to support the Internet, mobile computing, and every-increasing server deployments. Thus, the client-server computing model of the early

1990s will extend beyond local area networks (LAN) and private wide-area networks (WAN) to encompass Internet Web architectures and wireless communications.

The final component of the "GIS of the future" operating environment scalability. In reference to computer systems, scalability is the ability to support a diverse set of operating environments; from PC to mainframe. Thus, the GIS applications operating environment will be transparent to the end user. The typical user may use a personal computer that draws upon data from the mainframe store data repository, demographic information from a data vendor via the Internet, and utilize applications that are partially resident on the personal computer, workstations, and mainframes.

Why will such a complicated computer model be necessary? The primary reason is the trend toward utilizing shared enterprise-wide data resources. With resource sharing, comes the necessity for controlling the data access, as well as the integrity of the data resource and any other programs that access the same data. Retailers are trying to leverage the value of the vast quantities of data they gather by using data warehouse structures and data mining tools. Data-mining technology exemplifies the current state of enterprise-wide database access software. Simply stated, data mining provides a very powerful query ability that lets users ask open-ended questions. In a GIS environment, such queries would populate the client map dataset. In other words, all enterprise data could be made available for spatial analysis using GIS tools. Several companies have developed powerful tools to develop and maintain gigantic databases, often terabytes in size (a terabyte is 1,000,000 gigabytes) just as database vendors are developing the structures necessary to accommodate dataset of virtually unlimited size.

Building such databases involves three major development steps. First, data from multiple sources and in multiple formats are loaded into the data warehouse. In planning for this stage, the developers must take into account the sources of all data, how it has been processed, and clearly define what each data element means. Often this data is captured through OLAP or OLTP processes (On Line Analytic Processing/On Line Transaction Processing). The second stage of warehouse development is scrubbing the data to correct errors and integrate the datasets into a comprehensive, universally accessible store of information. In constructing the data warehouse, these historical problems in the data are corrected by merging redundant data, resolving conflicting data, and integrating data from incompatible sources. Of course, all of the security and data integrity procedures provided by modern database software is available to the data warehouse. Simultaneously, the third step is taking place; a metadata creator tracks and records the source of the data, saving the information necessary for users to understand the meaning of the data in the warehouse. Once the data warehouse is created, users can ask open-ended questions use drill-down tools to reorganize and summarize the data to satisfy their business criteria, and present the results of their analysis using GIS of other visualization and reporting tools. In effect, a retailer can determine which customers bought item "x" along with item "y" and their demographic profile. Armed with this information, the retailer could design promotional mailings to that specific customer or all potential customers with similar profiles located within 10 miles of the stores.

Such uses are rapidly becoming reality in today's retail environment. Databases are no longer just the basis for tactical activities of the marketing department nor is GIS strictly the domain of the real estate and market analysis departments. Wal-Mart recently announced its intent to triple the size of its data warehouse to 24 terabytes of storage capacity (*Computerworld*, Feb. 17, 1997, v31, n7, p8). This expenditure, estimated to be $40 to $60 million, will capture data at the individual transaction level to provide a clearer picture of customer buying habits, efficiency of promotions, and provide the basis for better store layouts.

What will this very complicated scenario mean in practical terms to location research? It means that data structures need not be limited to one physical location or even one company. Users can seamlessly integrate all available information pertinent to their project. High-end data analysis will be automated and available on an as-needed basis. Sales forecasting will be automated with systems that can provide explanations for its conclusions. New sites will be analyzed in terms of their impact on the remaining network of stores within a market. Complete real estate submittal packages will be prepared automatically for the real estate committee including complete pro forma analysis and multimedia presentation materials.

Such is the new world of location research. While some of these ideas may seem fantastic given the labor-intensive nature of current site location research practices, all of these ideas are currently in research and development or already available as working systems. It is our feeling here at Thompson Associates, that all of these ideas will become reality within the next decade. We are working hard to make it happen.

Resources

International Council of Shopping Centers
662 5th Avenue
New York, N.Y. 10022
(212) 421-8181

Urban Land Institute
1025 Thomas Jefferson Street, N.W.
Suite 500W
Washington, D.C. 20007
(202) 624-7000

Bureau of the Census
Resources Division
Room 1412-3
Washington, D.C. 20233
(301) 457-1722

Glossary

Chapter 1

Analog method. A research methodology that uses analogies and inferences derived from studying the performance of existing stores to forecast the performance of new stores.

Consumer research. The systematic gathering, analyzing, and interpreting of information about consumers—where they live, how they shop, what they buy, their demographic characteristics, and their perceptions.

Correlation analysis. A statistical technique for determining the degree of relationship, usually linear, that exists between two variables.

Enhanced analogs. A variation of basic analog methodology in which variables such as population, demographic characteristics, competition, access patterns, economic considerations, and trade area segments (rather than total trade areas) are used in the selection of analogous situations to forecast the performance of new stores.

Gravity models. A research technique used in forecasting sales for convenience-oriented retailers, in which store sizes and sales volumes, distance, and trade area expenditure potential are the main variables used to explain a store's sales potential.

Location research. The identification and quantification of factors that most significantly affect store sales performance, and the development and application of tools that aid in the identification of potential expansion or deployment opportunities.

Normal curves. Graphic devices to depict the relationship between a store's or chain's sales penetration levels and the distance that customers travel to shop at the store, under a variety of competitive and/or demographic scenarios.

Quantitative site selection techniques. Estimating the sales potential for proposed store sites by analyzing the sales patterns of customers from existing stores.

Regression analysis. A statistical technique for measuring the degree of association, usually linear, between and among several variables. The variables to be predicted are termed dependent and the predictive variables are termed independent.

Reilly's Law of Retail Gravitation. The probability of a consumer choosing a specific store at which to shop is a function of a store size (and its perceived selection) and distance to the store, relative to other stores and the distance to them. Reilly's Law defines the line of equal attractiveness between stores or market centers.

Spatial allocation models. Using generators of movement of retail dollars such as population and competitive retail locations, along with the restraints on such interaction, spatial allocation provides an equilibrium model format with which to view the retail environment. Once a balance model is achieved, a new outlet can be introduced to simulate the sales impact of the store.

Chapter 2

Consumer expenditure model. An estimate of total consumer expenditure potential for a type of store or product group, calculated by multiplying dollars spent (per-capita or per-household) by the density (people or households) of the relevant trade area being studied.

Location. A marketing term used to describe an area that may or may not represent a viable site for a retail store.

Market characteristics. Quantitative and qualitative aspects of a market area that may affect the well-being and applicability of the retail environment.

Measures of density. Data such as population, households, or businesses expressed as a ratio per unit of geography (such as population per square mile).

PCW. An estimate of per-capita weekly food store expenditure potential, based on demographic characteristics such as income, household size, age, and so on.

Primary data. Data collected firsthand as part of a research project to satisfy the needs of the investigation.

Secondary data. Data that have already been collected for some other purpose, but are usable in the current research investigation.

SIC. Standard Industrial Classification a U.S. government code whereby all businesses are categorized by type, size, and so forth.

Site. A real estate term referring to a physical parcel of land.

Site characteristics. Qualitative observations concerning visibility, parking, ingress/egress, and so on, used to describe a site's viability as a retail location.

Statistical validity. The degree to which research findings and conclusions accurately reflect the premises, inferences, or data from which they were drawn.

Trade area. A geographic area within which a store's influence is felt, and from which a store obtains a majority of its customers.

Trade area access. The ease or difficulty with which prospective shoppers travel from the place of origin to a store or shopping center.

Chapter 3

Block groups. Bureau of the Census-defined area.

Capture rate. The proportion of a store's sales volume that originates from a unit of geography such as its trade area or surrounding ZIP Codes, census tracts, and so forth.

Census tract. Bureau of the Census-defined area.

Centroid. Center of the population density for a unit of geography such as a trade area segment.

Competitive "Noise." Another term for indirect competitors—those competitors not having a direct impact on a retailer.

Geocoded. Identifying the location of an address on a map, or within a unit of geography such as ZIP Code, census tract, block group, etc., usually using proprietary computer software.

GIS. Geographic Information Systems.

Penetration level. A measurement of how effectively a store is serving the various geographic components of its trade area, usually expressed in terms of sales per unit of geography (e.g., ZIP Code, census tract).

POS. Point of Sale.

Sister stores. Other stores in the same chain.

Transient population. Vehicles that drive past a store on a daily basis.

Chapter 4

Target customer profile. The demographic/psychographic profile of those consumers who, absent differences in relative convenience or competition, result in the highest sales penetration level for a retail concept.

Chapter 5

Adjacent. A competitor that is located adjacent, or effectively adjacent, to the database store.

Impacting. A competitor that is equally, or almost equally, situated to serve a particular geography as the database store, that is, it does not enjoy a significant geographic advantage or disadvantage relative to the database store.

Intercepting. A competitor that is located in a position to intercept, or cut off, all or most of the consumers in a particular geography that would otherwise shop the database store.

Retail synergy. The presence or absence of other retailers in the immediate vicinity of the site that may attract consumer traffic to the area.

Chapter 6

Market prioritization analysis. Identifying markets that have the greatest concentration of in-profile customers and weakest competition.

Chapter 7

Holes. Concentration of in-profile consumers that has been overlooked because they are not adequately served by a mall or other defined retail format.

In-profile consumer. Consumer whose demographic characteristics closely match the profile of the ideal consumer for a specific retail store.

Chapter 8

Averaging. Developing a forecast for each trade area sector by averaging the sales penetration levels derived from the comparable analogs and applying this average to the trade area sector being analyzed.

Convenience goods. Everyday consumables such as groceries typically not compared at a number of other stores before the purchase.

Convergence. Focuses on the subgeography forecast of where the analog penetration levels cluster around a central value.

Macro-analogs. Take into account the entire trade area as a single entity.

Micro-analogs. Break the trade area into its component geographic parts.

Outliers (analog). Sets of data which fall outside of the norm (either high or low).

Shopping goods. Merchandise for which consumers typically comparison shop (i.e., furniture, new cars, fashion, etc.).

Chapter 9

Curve. An indication of how a store's sales are distributed over geography, in which the higher the curve, the greater the store's proportion of sales generated from consumers residing close to the store (also called **"pulling power"**).

Density radius. An indication of the geographic extent of a store's trade area.

Drawing power. The percentage of a store's sales derived from a specified geography such as trade area (also called **"percent explained"**).

Float. Sales potential dollars from the trade area being studied that are not absorbed by the stores in the trade area (also known as **"leakage"**).

Image (Flavor, Power). A number assigned by the gravity model to each store in the model as an indication of its market strength means of indicating its strength relative to all of the other modeled stores.

Leakage. Sales potential dollars within the area being studied that are not absorbed by the stores being studied (also known as **"float"**).

Percent explained. The percentage of a store's sales derived from a specified geography such as trade area (also called **"drawing power"**).

Pulling power. An indication of how a store's sales are distributed over geography, in which the higher the curve, the greater the store's proportion of sales generated from consumers residing close to the store (also called **"curve"**).

SLASH. The name of the original Super Valu gravity mode, an acronym of **Store Location Analysis System Heuristically**.

Chapter 10

Sales transfer analysis. The impact that a new store or stores will have on the existing store network.

Chapter 11

Aggregate forecasting. Developing a sales forecast for an entire trade area and not its sectors.

Black-box reliability. The inference that the results of a forecast are correct if the methodology is mathematically or otherwise accurate.

Disaggregate forecasting. Developing an independent forecast for each geographic component of a store's trade area and then aggregating the forecast to a trade-area level.

Forecast convergence. The use of two or more independent forecasting methodologies to generate a sales estimate for a proposed retail location, thereby improving the level of confidence in the estimate.

Ramp-up (maturity). The difference between the initial sales performance of a newly opened retailer and its ultimate sales performance once it is better known and accepted by area consumers.

Chapter 13

Share of space (share of markets, sales forecasting). This methodology assumes that the percentage of space that a store has will equal its proportion of the total market share that it will achieve. It is necessary to calculate the total size of all competitors serving the area being studied, add to that figure the size of the subject store, and find the percentage of the total that the subject store will have.

Start-up stores. Innovative retail facilities that distinguish themselves from other stores that sell comparable items due to unique their image, merchandise selection, specialization, pricing, and/or customer appeal.

Chapter 14

Customer exit survey. A more extensive survey than the customer source survey that solicits answers to a broader range of questions.

Customer source survey. A survey conducted with a sampling of a store's shoppers to determine their residence or place of work. Usually used to quantify the geographic draw of a store or entire shopping center so that a trade area can be defined.

Focus group studies. In-depth interview with small groups of consumers concentrating on issues that may or may not need further quantification.

Tagging. A graphic plotting of a customer's purchase and trip origin.

Chapter 15

CHAID. Chi Square Automatic Interaction Detection; a nonparametive statistical analysis system that provides decision-tree structures used for analysis of variable causality and interactions.

GPS. Global Positioning Systems; a system of stationary satellites that send signals that can be triangulated for the purpose of precise measurement of position on the earth's surface.

Metadata. Data about data—its source, currency, quality, and suitability for a given purpose.

Neural network. Computer software systems that are patterned after the functioning of the human brain.

OLAP. On-line Analytical Processing; software used to capture transactional data from on-line systems, for example, point of sale (POS) or inventory management systems to aggregate or process the data for further analysis by other systems.

SKU codes. Stock Keeping Unit codes (bar codes); the inventory numbers for merchandise.

TIGER. Topographically Integrated Geographic Encoding and Referencing; the road network and address system used by the U.S. Bureau of the Census for enumeration purposes.

Index